AF595056

Technology in Rehabilitative Interventions for Children: Challenges and Opportunities

Technology in Rehabilitative Interventions for Children: Challenges and Opportunities

Editors

Alessandro Antonietti
Daniela Traficante

MDPI • Basel • Beijing • Wuhan • Barcelona • Belgrade • Manchester • Tokyo • Cluj • Tianjin

Editors
Alessandro Antonietti
Department of Psychology
Catholic University
of Sacred Heart
Milan
Italy

Daniela Traficante
Department of Psychology
Catholic University
of Sacred Heart
Milan
Italy

Editorial Office
MDPI
St. Alban-Anlage 66
4052 Basel, Switzerland

This is a reprint of articles from the Special Issue published online in the open access journal *Children* (ISSN 2227-9067) (available at: www.mdpi.com/journal/children/special_issues/Technology_Rehabilitative_Challenges).

For citation purposes, cite each article independently as indicated on the article page online and as indicated below:

LastName, A.A.; LastName, B.B.; LastName, C.C. Article Title. *Journal Name* **Year**, *Volume Number*, Page Range.

ISBN 978-3-0365-4292-8 (Hbk)
ISBN 978-3-0365-4291-1 (PDF)

Contents

About the Editors

Alessandro Antonietti

Alessandro Antonietti is full professor of Cognitive Psychology and head of the Department of Psychology at the Catholic University of the Sacred Heart in Milano (Italy). He is also the coordinator of the Cognitive Psychology Lab and of the master in Developmental Cognitive Disabilities. He is a member of the board of various international and national journals and academic editor of PlosONe, as well as reviewer of many scientific journals. He carried out experimental studies about creativity, problem-solving, decision-making, mental imagery, and analogy. He investigated the role played by media in cognition. He is interested in the applications of cognitive issues in the field of instruction and rehabilitation. He devised tests to assess thinking skills and programs to train cognitive abilities.

Daniela Traficante

Daniela Traficante, PhD, is Associate Professor of Developmental and Educational Psychology at Catholic University of Sacred Heart of Milan, and is the Director of the SPAEE (Service of Learning and Educational Psychology in Childhood), a clinical centre of the Department of Psychology. Her research interests are reading processes in typical and atypical populations and the relationship between learning disorders and wellbeing experience in children. Moreover, she is interested in the assessement of cognitive development and reading skills, and carried out research projects on evidence-based treatment of learning disorders in literacy.

Preface to "Technology in Rehabilitative Interventions for Children: Challenges and Opportunities"

This Special Issue is aimed to offer an overview of studies presenting new rehabilitation approaches addressed to children with neurodevelopmental disorders, designed to enhance the effects of learning processes through the use of new technologies. Several studies showed that the application of new information, cybernetic, and electromedical technologies has provided the opportunity to affect, through different mechanisms, the neuronal plasticity in a more or less direct way. Moreover, the use of new technologies in rehabilitation played a main role during the COVID-19 pandemic, as many clinicians used telerehabilitation in order to keep in touch with their patients and going on with therapy planning. Now it might be relevant to consider what the true added value of technological instruments in rehabilitation is, irrespective to the emergency due to the pandemic.

The contributions of this Special Issue can offer some insightful data and advice on the use of technology in rehabilitation and telerehabilitation to researchers, rehabilitators and clinicians, and pratictioners (psychologists, neuropsychologists, speech therapist, etc.) working in the field of neurodevelopmental disorders. The authors of the papers come from some of the most valued Italian scientific institutions in the field: the Child Neuropsychiatry Unit of IRRCS Fondazione Policlinico "A. Gemelli" (Rome), the Developmental Neurology Unit of Foundation IRCCS Neurological Institute "C. Besta" (Milan), the Unit of Child Psychopathology of Scientific Institute IRCSS "E. Medea" (Bosisio Parini), and the Department of Psychology, Catholic University of Sacred Heart (Milan).

We hope that this Special Issue can provide an interesting overview on several questions raised by the application of technology to the treatments of neurodevelopmental disorders and contribute to distinguish appearances from the actual benefits of this approach.

Alessandro Antonietti and Daniela Traficante
Editors

 MDPI

Editorial
Technology in Rehabilitative Interventions for Children: Challenges and Opportunities

Daniela Traficante * and Alessandro Antonietti 

Department of Psychology, Catholic University of the Sacred Hearth, 20123 Milan, Italy; alessandro.antonietti@unicatt.it
* Correspondence: daniela.traficante@unicatt.it

Citation: Traficante, D.; Antonietti, A. Technology in Rehabilitative Interventions for Children: Challenges and Opportunities. *Children* **2022**, *9*, 598. https://doi.org/10.3390/children9050598

Received: 18 April 2022
Accepted: 20 April 2022
Published: 22 April 2022

Technology innovation has been leading to the development of an increasing number of applications that aim to support the rehabilitation of cognitive functions. What are the pros and cons of the use of such applications in training programs addressing children with neurodevelopmental disorders? The papers included in this Special Issue offer the opportunity to explore this topic from different perspectives as they are focused not only on the efficacy of interventions mediated by technology [1–3] but also on the impact that the use of technology in telerehabilitation, in particular during the COVID-19 pandemic, has had on wellbeing experience of children with Specific Learning Disorders and Cerebral Palsy [4]. Moreover, due to the relevance of the use of technology in the treatment of Specific Learning Disorders, Lorusso et al. [5] carried out a Delphi study to propose a set of best practices in order to guide the use of technology to realize training programs and to support learning processes in school activities.

The first question addressed by the Special Issue is the efficacy of the use of technology in the treatment of neurodevelopmental disorders. Lino et al.'s narrative review [1] presents an overview on the use of serious games using Virtual Reality (VR) and Augmented Reality (AR) to treat Developmental Coordination Disorder (DCD) in childhood. The authors argued that the "Internal Modeling Deficit" (IMD) characterizes the compromised motor ability of children with DCD. More precisely, the internal models of motor control are supposed to stem from two components: Mental Imagery (MI) and Action Observation (AO). Even though digital technology can offer the opportunity to support both these components, through the visualization and reproduction of motor patterns and strategies in children with DCD, few studies assessed the effectiveness of training based on VR/AR technologies. To date, there is preliminary evidence that such technologies can be beneficial due to the high level of children's engagement and enjoyment in attending treatment sessions with VR/AR, but further research is needed to obtain reliable data on their efficacy.

As for Specific Learning Disorders, Lorusso et al. [2] present their experience from the application of an automated training program, Tachidino, aimed at the treatment of reading and writing disorders. The software is based on the Visual Attention Training (AVG) and the Visual-Hemisphere-Specific Stimulation (VHSS) approach, which is grounded on Bakker's Balance Model. It allows therapists to tailor the duration and the flexibility of intervention in order to support the child's engagement and reading fluency. The assessment of the efficacy of the training program showed encouraging results, irrespective of the age of the children and the severity of the disease. Hence, such a training program can be considered a useful tool to promote reading ability in children with Dyslexia.

Serious videogames, such as the activities included in Tachidino, are worth considering because of their interactive modality. In addition, they fit for the implementation of telerehabilitation, which played a major role in supporting the wellbeing and the treatment of children with neurodevelopmental disorders during the COVID-19 pandemic. In fact, in that condition, technology offered therapists the opportunity to assure the rehabilitation program would continue. The unusual situation caused by the pandemic led researchers and practitioners to face a new question, i.e., the influence of the modality of the training implementation (online vs. in presence) on treatment efficacy. In order to give some clues on this issue, Cancer et al.'s study [3] compared the effects of Rhythmic Reading Training (RRT), realized under the direct supervision of the practitioner present, and the effects observed when the training was applied online under the remote control of the therapist to remediate reading deficits in children with Dyslexia. The results showed that there was not any difference between the two modalities. Therefore, this study can be considered a piece of evidence to support the use of telemedicine and telerehabilitation in the context of interventions addressed to pediatric populations.

The relevance of telerehabilitation in supporting the wellbeing experience in children with neurodevelopmental disorders is well documented by the contribution of Sarti et al. [4], who assessed the experience of children with Specific Learning Disorders and Cerebral Palsy during the pandemic period. The authors compared children who were involved in online treatment to children who did not have the opportunity to undertake telerehabilitation by analyzing a group of typically developing children (control group) as well. Children who were engaged in telerehabilitation showed a higher level of involvement in learning activities, with a higher level of perceived social support and respect, than the control group.

As the role of technology and telerehabilitation has been rising during the last decade and is likely to increase in the future, it might be important to identify the best practices that can provide professionals with suggestions and share experience among practitioners in order to offer effective support to the greatest number of children in any condition of implementation of the treatment. The paper by Lorusso et al. [5] presents a synthesis of a Delphy study conducted among Italian professionals, aiming to endorse some statements on the use of new technologies applied in the treatments of Dyslexia, which will be included within a European project on the use of VR and AR in learning to read. Overall, the respondents showed a positive attitude towards the use of Information and Communication Technology in rehabilitation programs and appreciated, in particular, the application of VR in training programs aimed at the automation of grapheme-to-phoneme conversion rules.

We currently have a large number of technological tools that can be employed in remediation interventions addressed to neurodevelopmental disorders at our disposal and the trend is constantly increasing, with new devices being and their implementation becoming widespread. What is crucial now is understanding the directions to be followed in order to orientate the efforts to improve the quality of rehabilitation thanks to technology towards relevant goals. To do so, the comprehension of the true added value of technological instruments is important. Technology is seductive because novelty generates expectations and enthusiasm, but what is new is not always better as well. A critical approach is needed to distinguish appearance from actual benefits. On the one hand, technology allows practitioners to optimize procedures, which indeed can be implemented through traditional means. In this case, the advantage produced by technology consists of saving time, improving precision, reducing costs, shortening the training periods of therapists, and so forth. On the other hand, technology allows professionals to do things that cannot be carried out with traditional tools, such as reaching new populations, involving parents in the treatment, combining different media and formats, and so on. The first kind of benefits must not be neglected, but obviously, the second kind is more intriguing and this seems to be the most interesting direction. In this perspective, the challenge is identifying the processes that underlie the application of technological devices: What changes across the rehabilitation treatment when certain tools are used? Do changes concern attitudes, motivation, feelings, cognition, communication, or social relations? Additionally, more specifically, which attitudes, motives, feelings, cognitive, communicative, and social processes are involved? We hope the Special Issue can provide readers some insights into this topic and suggest some possible future steps for research and intervention.

Author Contributions: All authors (D.T. and A.A.) contributed to write this editorial. All authors have read and agreed to the published version of the manuscript.

Funding: This research received no external funding.

Conflicts of Interest: The authors declare no conflict of interest.

References

1. Lino, F.; Arcangeli, V.; Chieffo, D.P.R. The Virtual Challenge: Virtual Reality Tools for Intervention in Children with Developmental Coordination Disorder. *Children* **2021**, *8*, 270. [CrossRef] [PubMed]
2. Lorusso, M.L.; Borasio, F.; Molteni, M. Remote Neuropsychological Intervention for Developmental Dyslexia with the Tachidino Platform: No Reduction in Effectiveness for Older nor for More Severely Impaired Children. *Children* **2022**, *9*, 71. [CrossRef] [PubMed]
3. Cancer, A.; Sarti, D.; De Salvatore, M.; Granocchio, E.; Chieffo, D.P.R.; Antonietti, A. Dyslexia Telerehabilitation during the COVID-19 Pandemic: Results of a Rhythm-Based Intervention for Reading. *Children* **2021**, *8*, 1011. [CrossRef] [PubMed]
4. Sarti, D.; De Salvatore, M.; Pagliano, E.; Granocchio, E.; Traficante, D.; Lombardi, E. Telerehabilitation and Wellbeing Experience in Children with Special Needs during the COVID-19 Pandemic. *Children* **2021**, *8*, 988. [CrossRef] [PubMed]
5. Lorusso, M.L.; Borasio, F.; Da Rold, M.; Martinuzzi, A. Towards Consensus on Good Practices for the Use of New Technologies for Intervention and Support in Developmental Dyslexia: A Delphi Study Conducted among Italian Specialized Professionals. *Children* **2021**, *8*, 1126. [CrossRef] [PubMed]

 *children*

Review

The Virtual Challenge: Virtual Reality Tools for Intervention in Children with Developmental Coordination Disorder

Federica Lino [1], Valentina Arcangeli [2,3] and Daniela Pia Rosaria Chieffo [2,3,4,*]

1 Clinical Psychology Unit, Memory Clinic, IRRCS Fondazione Policlinico A. Gemelli, 00168 Roma, Italy; federica.lino.psy@gmail.com
2 Clinical Psychology Unit, IRRCS Fondazione Policlinico A. Gemelli, 00168 Roma, Italy; valenti-na.arca@gmail.com
3 Child Neuropsychiatry Unit, IRRCS Fondazione Policlinico A. Gemelli, 00168 Roma, Italy
4 Faculty of Medicine and Surgery, Catholic University of Sacred Heart, 00168 Roma, Italy
* Correspondence: danielapiarosaria.chieffo@policlinicogemelli.it; Tel.: +6-015-3364 or +63-015-3364; Fax: +6-015-5676 or +63-015-5676

Abstract: This narrative review highlights the latest achievements in the field of tele-rehabilitation: Virtual Reality (VR) and Augmented Reality (AR) serious games aimed at restoring and improving cognitive functions could be effectively used in Developmental Coordination Disorder Training. Studies investigating the effects of the abovementioned tech applications on cognitive improvement have been considered, following a comprehensive literature search in the scientific electronic databases: Pubmed, Scopus, Plos One, ScienceDirect. This review investigates the effects of VR and AR in improving space/motor skills through mental images manipulation training in children with developmental coordination disorders. The results revealed that in spite of the spreading of technology, actually only four studies investigated the effects of VR/AR tools on mental images manipulation. This study highlights new, promising VR and AR based therapeutic opportunities for digital natives now available, emphasizing the advantages of using motivational reward-oriented tools, in a playful therapeutic environment. However, more research in this filed is needed to identify the most effective virtual tool set for clinical use.

Keywords: neurorehabilitation; cognitive enhancement in children; tech mediated rehabilitation; developmental coordination disorders; virtual reality

Citation: Lino, F.; Arcangeli, V.; Chieffo, D.P.R. The Virtual Challenge: Virtual Reality Tools for Intervention in Children with Developmental Coordination Disorder. *Children* **2021**, *8*, 270. https://doi.org/10.3390/children8040270

Academic Editors: Alessandro Antonietti and David E. Mandelbaum

Received: 30 December 2020
Accepted: 29 March 2021
Published: 1 April 2021

1. Introduction

In the last decades the number of tech applications for cognitive Rehabilitation in childhood has increased, thanks to technological innovation. Virtual Reality (VR) and Augmented Reality (AR) have arisen as a promising approach in the field of clinical rehabilitation. This narrative review aims at pointing out the most recent advances in the field of neuro-rehabilitation: It will focus on studies using VR and AR serious games in the treatment of Developmental Coordination Disorder (DCD) in childhood. Specifically, VR and AR training may be useful to enhance mental images manipulation such as rotation, spatial visualization, spatial orientation, and motor plan programming, offering children a constant feedback on their performance.

As a matter of fact, children with DCD show a reduced ability to imagine a motor act (esp. from a first-person perspective), which has been shown in research using mental limb rotation and visually guided pointing tasks [1]. Other works show that Mental Imagery (MI) ability is correlated with the ability to implement online corrections in healthy adults [2] and children with DCD [3].

Virtual environments have been found effective in representing further contexts for children to test their adaptive strategies and to improve responses. The present work has two advantages over previous works: First, it depicts the most recent state of the art of VR and AR application in children with DCD and, second, it identifies useful differences in terms of tools and methods applied to approach digital native's neurorehabilitation. The main goal of this work is giving a view of the studies carried out so far and encouraging both researchers and clinicians to identify what should be further investigated to increase knowledge in this field of study.

2. Developmental Coordination Disorder

As described in the American Psychiatric Association's (Washington, DC, USA) latest edition of the Diagnostic and Statistical Manual of Mental Disorders (DSM-5) [4], the child with DCD has motor coordination below expectations for his or her chronologic age and therefore may have been described as "clumsy" and may have had delays in early motor milestones, such as walking and crawling. Difficulties with coordination of either gross or fine motor movements, or both, could interfere with academic achievement or activities of daily living. Coordination difficulties do not relate to a medical condition or disease (e.g., cerebral palsy, muscular dystrophy, visual impairment, or intellectual disability). If intellectual disability is existing, the motor difficulties are in excess of those expected for the child's IQ. In the previous DSM edition (DSM-IV-TR) [5] DCD was included under the broad category of "learning disorders"; In DSM-5, it is subcategorized as a motor disorder within the broader category of "neurodevelopmental disorders." An additional criterion included in DSM-5 is that the onset of symptoms occurs during the developmental period.

All children with DCD show some impairment in motor learning and in new motor skill acquisition, as opposite to adult apraxia which is a disorder in the execution of already learned movements. There is no consensus on etiology of DCD. Intragroup approach through factor and cluster analysis highlighted that "motor impairment in DCD children varies both in severity and nature" [6]. Indeed, most studies have used screening measures of performance on some developmental milestones derived from global motor tests. Vaivre-Douret [6] has investigated in his review different functions together with standardized assessments, such as neuromuscular tone and soft signs, qualitative and quantitative measures related to gross and fine motor coordination and specific difficulties-academic, language, gnosic, visual motor/visual-perceptual, and attentional/executive in order to allow a better identification of DCD subtypes with diagnostic criteria and to provide an understanding of the mechanisms and of the cerebral involvement [7].

A systematic review of 16 studies involving school-aged children, published in 2011, showed significantly greater odds of developmental coordination disorder among children who had very low birth weights (<1500 g) or were very preterm (<32 week) than among age-matched controls born at term with normal birth weights (odds ratio [OR] 6.29, 95% confidence interval [CI] 4.37–9.05 to 8.66, 95% CI 3.40–22.07) [8]. Boys are 1.7 to 2.8 times more likely than girls to have the disorder [9].

In a prospective, population-based cohort study in southwestern England, probable developmental coordination disorder was identified in 346 children aged 7–9 [10]. Risk factors associated with the disorder were difficulties with attention (OR 1.94, 95% CI 1.17–3.24), social communication (OR 1.87, 95% CI 1.15–3.04), repetition of nonwords (e.g., unfamiliar spoken nonwords that the child is asked to repeat) (OR 1.83, 95% CI 1.26–2.66), spelling (OR 2.81, 95% CI 2.03–3.90), and writing and reading (OR 3.35, 95% CI 2.36–4.77) [10].

2.1. Conventional Treatment in DCD

A plausible hypothesis to explain the compromised motor ability of children with Developmental Coordination Disorder (DCD) suggests a substantial deficit in their ability to utilize internal models for motor control. A dysfunction in this mode of control is thought to compromise motor learning capabilities [1].

Children with DCD have problems generating and/or monitoring a mental (action) representation of intended actions, termed the "internal modeling deficit" (IMD) hypothesis [11].

Internal modeling deficits (IMDs) have been proposed as a neurological cause of DCD [1]. According to the IMD hypothesis, the sensory-motor integration in the internal model is dysfunctional in children with DCD, which reduces their ability to use predictive motor control [1,12]. According to Nobusako [13] "before slow, sensory-motor feedback becomes available, internal models provide stability to the motor system by predicting the outcome of movements. This allows rapid online correction".

Two approaches target the Internal Modeling: Mental Imagery (MI) and Action Observation (AO). According to Adams [14], the above mentioned are two faces of the same coin. Several studies combining MI and AO seem promising [15,16], confirming that MI, in Adam's words, is "a prime modality that may serve motor intervention for children's motor problems". According to Marshall [17], an integrated MI/AO program through digital technology would be more promising, taking into account that children with DCD have difficulties in programming strategies by themselves only, while no difficulty is proved in learning once they have observed the correct strategy.

Other studies [18] showed that mental rotation (MR) ability is implicated in the successful execution of a motor task. This finding leads to the necessity of programming strategies and tools pointing to work on mental imagery in the treatment of DCD.

2.2. *Mental Imagery*

A mental image is a product of cognitive activity which enables to represent reality through recall, manipulation, reproduction of objects and events without sensorial stimulation [19]. People can experience mental images in all sensorial modalities. Therefore, if we can perceive in an auditorily or olfactorily way, "we can also have auditory, olfactory, tactile mental imagery" [20]. Nevertheless, the most common sensory modality through which we experience MI is vision [21]. Motor MI refers to a reproduction of a motor sequence of movements without perceptually witnessing that movement.

MI is not a single ability but emerges from a crossroads of singles abilities. Working memory has an important role in MI. According to Kosslyn [22], in order to generate mental images, a recall of images from long term memory to working memory is necessary. The image is then put into the "visuo-spatial-sketch pad" [23]. Working memory plays a central role in mental rotation tasks.

The same neuronal mechanisms underpinning simulation (imagery) are involved in real execution of actions [24]. As an evidence in visual MI, which is the most studied, earliest visual motor cortex (areas 17 and 18) is involved [25]. Other studies [18] have shown that mental rotation and motor performance tasks may share a similar subprocess.

MI have been investigated to examine the cognitive aspects linked to action and movement control. The main advantage in using MI in rehabilitation trainings is the possibility to significantly increase the number of task repetitions since we use a mental recall of motor tasks. Wilson [26] pioneered studies investigating the direct impact of MI training on DCD providing interesting results. Later, Adams [14] demonstrated the theoretically principled protocol for MI training in DCD. By using Internal Modelling of movements, the child is facilitated in predicting the consequences of movements. This is possible because, during the training, he has acquired information on his internal feeling of the movement to make predictions on movement outcomes.

2.3. *Technology in DCD Rehabilitation*

In the last decades, many tech tools enabling motor skills treatment in Neurodevelopmental Disorders and in the field of DCD were developed. This has been possible because of a growing interest in tailored tools, able to meet patient specific needs. In such a framework, technological tools emerge as motivating and stimulating devices, adaptable to a single child's needs.

Telemedicine is currently developing in Italy and the National Healthcare System (NHS) has not exploited all the possibilities it offers yet [27]. In the field of Pediatric Medicine, telemedicine has the advantage of providing care and training in a non-medical environment [28,29]. Moreover, telemedicine allows custom-made training procedures, making daily interventions possible when needed. Research has contributed to a better understanding of the process underpinning children compliance to treatment. A game-like training setting has proved to be one of the most effective features for children in terms of motivation to the treatment.

In the last 20 years, gaming industry flourished and there was a combination between electronic games and neurorehabilitation research. "Serious Games" were born from this combination. With the term Serious Games we are referring to games whose peculiarity is not mere entertainment, but the empowerment of cognitive and motor function [30]. Video games training for rehabilitative purpose has been widely validated both in motor rehabilitation [31] and in cognitive empowerment and rehabilitation in several disorders [32–34] ensuring a similar efficacy compared to conventional treatments.

Regarding VR Intervention, Wilson [35] distinguishes between off the shelf tools (such as Wii fit) and specifically designed tools for rehabilitation (as Tele Rehab). Several studies have shown the effectiveness of off the shelf tools, combined with the convenience in terms of costs and usability: Nintendo Wii fit [36,37], Sony's Playstation Eye Toy [38], Xbox 360 Kinect, and Playstation 3 [39].

Other studies explored the advantages of a specific design for tools in rehabilitation such as the AR Serious Game "Athynos" [40], an AR tool specifically developed to target cognitive/motor functions in Dispraxia.

2.4. *Virtual Reality and Augmented Reality in DCD Treatment*

Among various technologies explored to work on DCD, one recent and largely unexplored technology is Serious Games combined with VR and AR.

VR is defined as a three dimensional immersive and interactive experience occurring in real time [41]. AR can be described as a real environment which is 'augmented' by means of virtual objects through the use of computer graphic technology. Compared to VR "in which the users while immersed cannot see the real world, AR allows the user to see the real world with virtual objects superimposed upon or composited with the real world" [42].

VR has been successfully employed in the neurorehabilitation of several disease in adulthood after brain injury or stroke [43–46] and in the case of neurodegenerative diseases [47,48]. In childhood, VR and AR have been found effective in several conditions such as Autism Spectrum Disorder [49,50], Attention Deficit Hyperactivity Disorder (ADHD) [51–54], or cerebral palsy [55–58].

AR has been found effective in empowering coordination skills in children. A study by Avila Pesantez [40] investigated the effects of an AR Training using a specifically designed tool called Athynos. Athynos was a prototype designed according to practice standards proposed by experts in the field of Dispraxia. The objectives inspiring designers were the improvement of hand-eye coordination skills, feedback, interactivity and problem solving.

Despite the good chances offered by these technologies, only few studies applied VR and AR to DCDs. In 1987, Mc Clurg [59] proved that tridimensional object manipulation would lead to visuospatial skills improvement. In the 90's Mc Comas [60] investigated the generalization effects of the improvement in visuospatial abilities gained through VR, confirming the generalization of the effects outside of the VR environment. The abovementioned results need to be further investigated [61] especially with regard to the extension of the generalization effect to more complex tasks.

In his study, Wilson [35] highlighted that VR could have an impact on different dimensions considered by the International Classification of Functioning [62]: Level of impairment, activity performance and skills, Participation, environment, personal factors (such as motivation or interests). With regard to training oriented to the empowerment of motor programming skills and mental rotation skills, VR enables children to manipulate 3D objects having an immediate feedback on task success into a realistic context. This represents one of most important advantages of VR technology in ecological terms.

3. Materials and Methods

This narrative review was implemented on the basis of a research conducted on major scientific online databases (PubMed, Scopus, Plos One, ScienceDirect). We included randomized controlled trials, reviews, systematic reviews, metanalysis, book chapters, and official guidelines from scientific associations published until December 2020. Our literature search was based on the following search terms: "Developmental Coordination Disorder" or "Coordination Disorder in children" or "motor coordination skills" or "mental rotation skills" or "mental images" or "mental images manipulation" combined with "VR treatment" or "VR/AR rehabilitation" or "VR intervention" or "Virtual remediation" or "Virtual Reality" or "Augmented Reality" or "virtual Reality training".

To be considered for this review studies had to meet the following criteria. First, the study aim had to examine the impact of a VR or AR training on DCD in children; Second, the study protocol had to implement mental images manipulation such as rotation, spatial visualization, spatial orientation, and motor plan programming. Single case studies were not included in this study because of the lack of generality of obtained effects.

4. Results

From the literature search, only four studies specifically investigated the potential of VR or AR technology in DCD rehabilitation exploring mental images manipulation improvement.

In 2011, Straker [63] investigated the impact of a new VR video game on motor coordination skills in children with DCD. The study sample consisted of 30 children between the age of 10–12 with poor motor coordination ability (<15th percentile). Children underwent 16 weeks training in two different conditions: 'Non active computer games' and 'active computer games'. Basic assumption was that children spend nowadays many hours playing video games and this reduces time spent in motor activities (with a consequent impact on motor competency). The first outcome measure was motor coordination ability, assessed by kinematic and kinetic three-dimensional motion analysis laboratory measures and two advanced mechanical technology force plates; Secondly, physical activity and sedentariness assessed through accelerometry have been considered; Coordination in daily life was measured by parent report questionnaire; Self-efficacy, anxiety, and mood outcomes were evaluated by self-report scales. The researchers' hypothesis was that a change in the nature of movement of children playing computer games would lead to a beneficial outcome on their motor and physical ability and sense of confidence. The results of the study did not give and exhaustive answer on the matter of research.

In 2013, Ashkenazi [38] investigated a low-cost VR intervention program for children with DCD. The technological off-the-shelf VR tool selected for this study was the Playstation R 2 Eye Toy system. It consists of a camera for 2D gesture recognition which enables virtual object manipulation on a screen. A visual and acoustic signal provided an immediate feedback for success in the 8 games/tasks proposed. The small sample for this study consisted of 9 children between 4–6 years old. Subjects underwent a weekly 1 h training for 12 weeks. Trainings targeted motor planning, balance control, eye-hand coordination, and multitasking. A baseline and follow-up recording were obtained. Despite some weaknesses, such as the small sample and the lack of a control group, statistically significant results in the primary outcome measures-namely, Movement Assessment Battery for Children-2 (M-ABC-2) and DCD Questionnaire (DCD-Q)-emerged. Even though motivation was not directly measured in this study, authors pointed out the children's enthusiasm for the device, supported by children preference for game-like training activities. For this reason, future studies could explore this feature in depth measuring variables such as motivation and enjoyment.

In 2020, a randomized controlled study [64] investigated the impact of a VR Serious Game training on motor control on a sample of 40 children with DCD, aged between 7 and 10 years. Specifically, the training targeted predictive control, described by the authors as the internal modelling of action in terms of MI and motor planning. The training consisted of 16

30 min session of Xbox 360 Kinect games administrated over an 8 week period. The selected VR device was unfamiliar to all children. Tasks to empower internal model control consisted in object manipulation and object control. Results demonstrated long lasting improvements, verified in the follow up phase. The outcome measures were MI skills (assessed through mental rotation task), action planning, and rapid and online control skills.

AR potentialities have also been investigated in the already mentioned study by Avila Pesantez [40], who compared two therapy methods (manual puzzle and AR) in 40 children (50% male and 50% female of average age 7.3 years). At each session the time of activities execution and the quality of performance for each student were registered. The study found significant difference between the two methods. The AR Serious game Athynos showed statistically significant gains in performance, described as less time spent on activities execution and less time spent in training, compared to the manual puzzle therapy. Improvement in motor level and hand and eye coordination were deducted from a boxplot of performance. There was no reference for a specific motor eye coordination assessment. Anthynos exploited the possibilities given by technology: It was programmed to give to children a constant feedback on achievements and tasks success rate. Lastly, the study pointed out another benefit of a similar approach in rehabilitation: Engaging children in proposed activities (taking into account that they are digital natives) ensuring as a consequence a good treatment compliance.

5. Discussion

The purpose of the present narrative review was twofold. First, we aimed to provide an up-to-date overview of the findings from VR/AR studies using mental images manipulation in DCD treatment for children. Second, we wanted to recommend researchers and clinicians worthy points to be additionally investigated in future research. Today, despite of the increasing use of technology in the clinical field, only four studies have been found interested in the role of VR and AR technologies in improving mental images manipulation for DCD treatment in children. Notwithstanding the literature is at a very early stage, some preliminary qualitative data can be recollected from the reviewed studies.

Results emerging from the presented studies, despite limitations, reveal a good potential in DCD treatment. Three out of the four studies presented in this review reported significant results in the outcome measures [38,40,64].

Straker [63] grounded his study approach on changing children' habits involved in playing computer games. To provide positive gains in motor control, the study assumed that a change from non-active to active video games playing could provide a gain in motor performance. Avila Pesantez [40] and colleagues, differently from the other studies, designed their own AR tool exploiting the motivating power of web graphic design and gratifying children with colorful avatars. Two out of the four studies [38,64] investigated the effectiveness of a training designed through an off-the-shelf tool (PlayStation R and Xbox) showing that low-cost tools can be exploited in clinical treatment of DCD in children. Because of the use of commercial virtual gaming tools, rehabilitation experience could be extended out of the hospital setting.

All the studies included in the present work diverged in many methodological aspects, such as sample size, study timeline, type of tools designed for intervention. These differences limited to some extent the chance to determine a direct link between a specific intervention and different outcomes. The limitation of the studies included in the present review are mostly the sample exiguity or methodological differences. However, the above-mentioned limitations are very common in pioneering studies exploring the use of new technological tools for neurorehabilitation.

To overcome this problem, the scientific community has been invited [61] to standardize methodologies in research for intervention in DCD through the use of the Template for Intervention Description and replication checklist by Hoffman [65]. This way better study procedures reporting will be possible and, consequently, a study comparison and replication will be possible. Randomized, controlled clinical studies with a larger sample are needed to give more accurate answers on effectiveness. At the current state of knowledge, well-structured VR and AR training programs implementing mental images manipulations could offer gains in terms of motor performance and perceived effectiveness.

Telemedicine is still under development in Italy but, since good results have been achieved from initial research, a great effort is required for researchers in this field of study. The potential of the use of VR/AR technology for children stands in the possibility of giving professional, daily treatment, intrinsically eliciting motivation and enjoyment. As a second advantage to be considered, children could be involved in well-structured game-like trainings outside of the hospital context.

Unfortunately, it could not be assessed which intervention is, among the others, more effective for a specific age-range. This was due to the fact that the studies investigated the efficacy of VR ad AR applications in different populations aged between four and twelve years (a wide age span). Greater attention should be dedicated in research to define whether a window of sensitivity is available or has to be agreed for this kind of treatments, since different age ranges have been considered in clinical studies conducted so far. Additionally, some attention should be dedicated to the longitudinal effects of virtual training in terms of improvement duration and outcome measures considered through life span. As an example, social involvement in children who underwent a DCD training during adolescence could be investigated.

The present review had some limitations. First, the goal of this work was to provide an up-to-date narrative overview of recent findings on the use of VR and AR Serious games training in DCD. To do this, we limited our search to a small amount of studies investigating the role of mental images on DCD rehabilitation. In addition, we considered only papers in English and we not considered single case studies because of the lack of generalizability of the outcomes. Moreover, as previously discussed, some methodological aspects of the reviewed studies also limited our findings (small sample size and different age range considered for intervention). Lastly, as is common knowledge, retention of learning skills for a certain time is a critical issue of any treatment. Future research on VR/AR interventions on DCD should include follow-up visits to evaluate the maintenance or the evolution of the benefits achieved over time.

6. Conclusions

VR and AR serious games training for DCD treatment are motivating tools, as they offer immediate feedback in very realistic game environments. Because of their features, VR and AR are among the most interesting tools researchers should look at to deliver tailored intervention for child's needs, as required by recent scientific guidelines [61].

The present review suggests that VR and AR interventions could have some beneficial effects in treating DCD. This work encourages future trials to consider mental images manipulation through VR/AR tool as a possible target for interventions on children with DCD.

More research is needed in this field: First, to explore further chances offered by this technology as a strategic asset in the field of DCD and, second, to identify procedures and targets that best fit the needs in this specific clinical area. This approach could lead to soon determine best practices for using VR and AR technology in this field of study.

Author Contributions: Conceptualization, methodology, investigation, resources, writing—original draft preparation, writing—review and editing, visualization, project administration, F.L., V.A., D.P.R.C.; All authors have read and agreed to the published version of the manuscript.

Funding: This research received no external funding.

Institutional Review Board Statement: Not applicable.

Informed Consent Statement: Not applicable.

Conflicts of Interest: The authors declare no conflict of interest.

References

1. Adams, I.L.; Lust, J.M.; Wilson, P.H.; Steenbergen, B. Compromised motor control in children with DCD: A deficit in the in-ternal model?—A systematic review. *Neurosci. Biobehav. Rev.* **2014**, *47*, 225–244. [CrossRef]
2. Hyde, C.; Wilmut, K.; Fuelscher, I.; Williams, J. Does implicit motor imagery ability predict reaching correction efficiency? A test of recent models of human motor control. *J. Mot. Behav.* **2013**, *45*, 259–269. [CrossRef] [PubMed]
3. Fuelscher, I.; Williams, J.; Enticott, P.G.; Hyde, C. Reduced motor imagery efficiency is associated with online control difficulties in children with probable developmental coordination disorder. *Res. Dev. Disabil.* **2015**, *45–46*, 239–252. [CrossRef]
4. American Psychiatric Association. *Diagnostic and Statistical Manual of Mental Disorders*, 5th ed.; American Psychiatric Publishing: Arlington, VA, USA, 2013.
5. American Psychiatric Association. *DSM-IV TR Diagnostic and Statistical Manual of Mental Disorders*; American Psychiatric Publishing: Washington, DC, USA, 2000.
6. Vaivre-Douret, L. Developmental coordination disorders: State of art. *Neurophysiol. Clin.* **2014**, *44*, 13–23. [CrossRef] [PubMed]
7. Vaivre-Douret, L.; Lalanne, C.; Golse, B. Developmental Coordination Disorder, An Umbrella Term for Motor Impairments in Children: Nature and Co-Morbid Disorders. *Front. Psychol.* **2016**, *7*, 502. [CrossRef]
8. Edwards, J.; Berube, M.; Erlandson, K.; Haug, S.; Johnstone, H.; Meagher, M.; Sarkodee-Adoo, S.; Zwicker, J.G. Developmental coordination disorder in school-aged children born very preterm and/or at very low birth weight: A systematic review. *J. Dev. Behav. Pediatr.* **2011**, *32*, 678–687. [CrossRef]
9. Larsen, R.F.; Mortensen, L.H.; Martinusson, T.; Andersen, A.-M.N. Determinants of developmental coordination disorder in 7-year-old children: A study of children in the Danish National Birth Cohort. *Dev. Med. Child Neurol.* **2013**, *55*, 1016–1022. [CrossRef]
10. Lingam, R.; Golding, J.; Jongmans, M.J.; Hunt, L.P.; Ellis, M.; Emond, A. The association between developmental coordination disorder and other developmental traits. *Pediatrics* **2010**, *126*, e1109–e1118. [CrossRef]
11. Gabbard, C.; Bobbio, T. The inability to mentally represent action may be associated with performance deficits in children with developmental coordination disorder. *Int. J. Neurosci.* **2011**, *121*, 113–120. [CrossRef]
12. Wilson, P.H.; Ruddock, S.; Smits-Engelsman, B.; Polatajko, H.; Blank, R. Understanding performance deficits in developmental coordination disorder: A meta-analysis of recent research. *Dev. Med. Child Neurol.* **2013**, *55*, 217–228. [CrossRef]
13. Nobusako, S.; Sakai, A.; Tsujimoto, T.; Shuto, T.; Nishi, Y.; Asano, D.; Furukawa, E.; Zama, T.; Osumi, M.; Shimada, S.; et al. Deficits in Visuo-Motor Temporal Integration Impacts Manual Dexterity in Probable Developmental Coordination Disorder. *Front. Neurol.* **2018**, *9*, 114. [CrossRef] [PubMed]
14. Adams, I.L.J.; Smits-Engelsman, B.; Lust, J.M.; Wilson, P.H.; Steenbergen, B. Feasibility of Motor Imagery Training for Children with Developmental Coordination Disorder—A Pilot Study. *Front. Psychol.* **2017**, *8*, 1271. [CrossRef] [PubMed]
15. Wilson, P.H.; Adams, I.L.; Caeyenberghs, K.; Thomas, P.; Smits-Engelsman, B.; Steenbergen, B. Motor imagery training enhances motor skill in children with DCD: A replication study. *Res. Dev. Disabil.* **2016**, *57*, 54–62. [CrossRef]
16. Buccino, G.; Arisi, D.; Gough, P.; Aprile, D.; Ferri, C.; Serotti, L.; Tiberti, A.; Fazzi, E. Improving upper limb motor functions through action observation treatment: A pilot study in children with cerebral palsy. *Dev. Med. Child Neurol.* **2012**, *54*, 822–828. [CrossRef]
17. Marshall, B.; Wright, D.J.; Holmes, P.S.; Williams, J.; Wood, G. Combined action observation and motor imagery facilitates visuomotor adaptation in children with developmental coordination disorder. *Res. Dev. Disabil.* **2020**, *98*, 103570. [CrossRef]
18. Hoyek, N.; Champely, S.; Collet, C.; Fargier, P.; Guillot, A. Is Mental Rotation Ability a Predictor of Success for Motor Performance? *J. Cogn. Dev.* **2014**, *15*, 495–505. [CrossRef]
19. Risoli, A.; Antonietti, A. *Il Corpo al Centro. Dalla Teoria alla Riabilitazione con il Metodo SaM, Edizioni Universitarie*; LED: Milano, Italy, 2015.
20. Nanay, B. Sensory Substitution and Multimodal Mental Imagery. *Perception* **2017**, *46*, 1014–1026. [CrossRef]
21. Kosslyn, S.M.; Seger, C.; Pani, J.R.; Hillger, L.A. When is imagery used in everyday life? A diary study. *J. Ment. Imag.* **1990**, *14*, 131–152.
22. Kosslyn, S.M.; Reiser, B.J.; Farah, M.J.; Fliegel, S.L. Generating visual images: Units and relations. *J. Exp. Psychol. Gen.* **1983**, *112*, 278–303. [CrossRef]
23. Baddeley, A.D. *Working Memory*; Oxford University Press; Clarendon Press: Oxford, UK, 1986.
24. Decety, J.; Jeannerod, M. Mentally simulated movements in virtual reality: Does Fitts's law hold in motor imagery? *Behav. Brain Res.* **1995**, *72*, 127–134. [CrossRef]
25. Kosslyn, S.M.; Pascual-Leone, A.; Felician, O.; Camposano, S.; Keenan, J.P.; Thompson, W.L.; Ganis, G.; Sukel, K.E.; Alpert, N.M. The role of area 17 in visual imagery: Convergent evidence from PET and rTMS. *Science* **1999**, *284*, 167–170. [CrossRef]
26. Wilson, P.H. Practitioner review: Approaches to assessment and treatment of children with DCD: An evaluative review. *J. Child Psychol. Psychiatry* **2005**, *46*, 806–823. [CrossRef]
27. Maresca, G.; Maggio, M.G.; De Luca, R.; Manuli, A.; Tonin, P.; Pignolo, L.; Calabrò, R.S. Tele-Neuro-Rehabilitation in Italy: State of the Art and Future Perspectives. *Front. Neurol.* **2020**, *11*, 563375. [CrossRef]
28. Schmeler, M.R.; Schein, R.M.; McCue, M.; Betz, K. Telerehabilitation clinical and vocational applications for assistive technology: Research, opportunities, and challenges. *Int. J. Telerehabil.* **2009**, *1*, 59–72. [CrossRef] [PubMed]
29. Zampolini, M.; Todeschini, E.; Bernabeu Guitart, M.; Hermens, H.; Ilsbroukx, S.; Macellari, V.; Magni, R.; Rogante, M.; Scattareggia Marchese, S.; Vollenbroek, M.; et al. Tele-rehabilitation: Present and future. *Ann. Ist. Super. Sanita* **2008**, *44*, 125–134. [PubMed]

30. Deutsch, J.E.; Westcott McCoy, S. Virtual Reality and Serious Games in Neurorehabilitation of Children and Adults: Prevention, Plasticity, and Participation. *Pediatr. Phys. Ther.* **2017**, *29* (Suppl. 3), S23–S36. [CrossRef] [PubMed]
31. Bonnechère, B.; Jansen, B.; Omelina, L.; Van Sint Jan, S. The use of commercial video games in rehabilitation: A systematic review. *Int. J. Rehabil. Res.* **2016**, *39*, 277–290. [CrossRef]
32. Franceschini, S.; Gori, S.; Ruffino, M.; Viola, S.; Molteni, M.; Facoetti, A. Action video games make dyslexic children read better. *Curr. Biol.* **2013**, *23*, 462–466. [CrossRef]
33. Clemenson, G.D.; Stark, C.E. Virtual Environmental Enrichment through Video Games Improves Hippocampal-Associated Memory. *J. Neurosci.* **2015**, *35*, 16116–16125. [CrossRef] [PubMed]
34. Jung, H.T.; Daneault, J.F.; Nanglo, T.; Lee, H.; Kim, B.; Kim, Y.; Lee, S.I. Effectiveness of a Serious Game for Cognitive Training in Chronic Stroke Survivors with Mild-to-Moderate Cognitive Impairment: A Pilot Randomized Controlled Trial. *Appl. Sci.* **2020**, *10*, 6703. [CrossRef]
35. Wilson, P.; Green, D.; Caeyenberghs, K.; Karen Steenbergen, B.; Duckworth, J. Integrating New Technologies into the Treatment of CP and DCD. *Curr. Dev. Disord. Rep.* **2016**, *3*, 138–151. [CrossRef]
36. Ferguson, G.D.; Jelsma, D.; Jelsma, J.; Smits-Engelsman, B.C. The efficacy of two task-orientated interventions for children with Developmental Coordination Disorder: Neuromotor Task Training and Nintendo Wii Fit Training. *Res. Dev. Disabil.* **2013**, *34*, 2449–2461. [CrossRef]
37. Hammond, J.; Jones, V.; Hill, E.L.; Green, D.; Male, I. An investigation of the impact of regular use of the Wii Fit to improve motor and psychosocial outcomes in children with movement difficulties: A pilot study. *Child Care Health Dev.* **2014**, *40*, 165–175. [CrossRef] [PubMed]
38. Ashkenazi, T.; Weiss, P.L.; Orian, D.; Laufer, Y. Low-cost virtual reality intervention program for children with developmental coordination disorder: A pilot feasibility study. *Pediatr. Phys. Ther.* **2013**, *25*, 467–473. [CrossRef]
39. Straker, L.; Howie, E.; Smith, A.; Jensen, L.; Piek, J.; Campbell, A. A crossover randomised and controlled trial of the impact of active video games on motor coordination and perceptions of physical ability in children at risk of Developmental Coordination Disorder. *Hum. Mov. Sci.* **2015**, *42*, 146–160. [CrossRef]
40. Avila-Pesantez, D.; Vaca-Cardenas, L.; Rivera, L.A.; Zuniga, L.; Miriam Avila, L. ATHYNOS: Helping Children with Dyspraxia Through an Augmented Reality Serious Game. In Proceedings of the 2018 International Conference on eDemocracy & eGovernment (ICEDEG), Ambato, Ecuador, 4–6 April 2018; pp. 286–290.
41. Pimentel, K.; Teixeira, K. *Virtual Reality: Through the New Looking Glass*; Windcrest/McGraw-Hill/TAB Books; Blue Ridge Summit: Franklin County, PA, USA, 1993.
42. Azuma, R.T. A survey of augmented reality. *Presence Teleoperators Virtual Environ.* **1997**, *6*, 355–385. [CrossRef]
43. Rose, F.D.; Brooks, B.M.; Rizzo, A.A. Virtual reality in brain damage rehabilitation: Review. *Cyberpsychol. Behav.* **2005**, *8*, 241–262, 263–271. [CrossRef]
44. Corbetta, D.; Imeri, F.; Gatti, R. Rehabilitation that incorporates virtual reality is more effective than standard rehabilitation for improving walking speed, balance and mobility after stroke: A systematic review. *J. Physiother.* **2015**, *61*, 117–124. [CrossRef]
45. Zanier, E.R.; Zoerle, T.; Di Lernia, D.; Riva, G. Virtual Reality for Traumatic Brain Injury. *Front. Neurol.* **2018**, *9*, 345. [CrossRef]
46. Jack, D.; Boian, R.; Merians, A.S.; Tremaine, M.; Burdea, G.C.; Adamovich, S.V.; Recce, M.; Poizner, H. Virtual reality-enhanced stroke rehabilitation. *IEEE Trans. Neural Syst. Rehabil. Eng.* **2001**, *9*, 308–318. [CrossRef]
47. Lei, C.; Sunzi, K.; Dai, F.; Liu, X.; Wang, Y.; Zhang, B.; He, L.; Ju, M. Effects of virtual reality rehabilitation training on gait and balance in patients with Parkinson's disease: A systematic review. *PLoS ONE* **2019**, *14*, e0224819. [CrossRef]
48. Maggio, M.G.; Russo, M.; Cuzzola, M.F.; Destro, M.; La Rosa, G.; Molonia, F.; Bramanti, P.; Lombardo, G.; De Luca, R.; Calabrò, R.S. Virtual reality in multiple sclerosis rehabilitation: A review on cognitive and motor outcomes. *J. Clin. Neurosci.* **2019**, *65*, 106–111. [CrossRef]
49. Yuan, S.N.V.; Ip, H.H.S. Using virtual reality to train emotional and social skills in children with autism spectrum disorder. *Lond. J. Prim. Care* **2018**, *10*, 110–112. [CrossRef]
50. Bellani, M.; Fornasari, L.; Chittaro, L.; Brambilla, P. Virtual reality in autism: State of the art. *Epidemiol. Psychiatr. Sci.* **2011**, *20*, 235–238. [CrossRef]
51. Crepaldi, M.; Colombo, V.; Mottura, S.; Baldassini, D.; Sacco, M.; Cancer, A.; Antonietti, A. Antonyms: A Computer Game to Improve Inhibitory Control of Impulsivity in Children with Attention Deficit/Hyperactivity Disorder (ADHD). *Information* **2020**, *11*, 230. [CrossRef]
52. Crepaldi, M.; Colombo, V.; Mottura, S.; Baldassini, D.; Sacco, M.; Cancer, A.; Antonietti, A. The Use of a Serious Game to Assess Inhibition Mechanisms in Children. *Front. Comput. Sci.* **2020**, *2*, 34. [CrossRef]
53. Bashiri, A.; Ghazisaeedi, M.; Shahmoradi, L. The opportunities of virtual reality in the rehabilitation of children with attention deficit hyperactivity disorder: A literature review. *Korean J. Pediatr.* **2017**, *60*, 337–343. [CrossRef]
54. Shema-Shiratzky, S.; Brozgol, M.; Cornejo-Thumm, P.; Geva-Dayan, K.; Rotstein, M.; Leitner, Y.; Hausdorff, J.M.; Mirelman, A. Virtual reality training to enhance behavior and cognitive function among children with attention-deficit/hyperactivity disorder: Brief report. *Dev. Neurorehabil.* **2019**, *22*, 431–436. [CrossRef]
55. Golomb, M.R.; McDonald, B.C.; Warden, S.J.; Yonkman, J.; Saykin, A.J.; Shirley, B.; Huber, M.; Rabin, B.; Abdelbaky, M.; Nwosu, M.E.; et al. In-home virtual reality videogame telerehabilitation in adolescents with hemiplegic cerebral palsy. *Arch. Phys. Med. Rehabil.* **2010**, *91*, 1–8.e1. [CrossRef]

56. Snider, L.; Majnemer, A.; Darsaklis, V. Virtual reality as a therapeutic modality for children with cerebral palsy. *Dev. Neurorehabil.* **2010**, *13*, 120–128. [CrossRef]
57. Ravi, D.K.; Kumar, N.; Singhi, P. Effectiveness of virtual reality rehabilitation for children and adolescents with cerebral palsy: An updated evidence-based systematic review. *Physiotherapy* **2017**, *103*, 245–258. [CrossRef]
58. Chen, Y.; Fanchiang, H.D.; Howard, A. Effectiveness of Virtual Reality in Children with Cerebral Palsy: A Systematic Review and Meta-Analysis of Randomized Controlled Trials. *Phys. Ther.* **2018**, *98*, 63–77. [CrossRef]
59. McClurg, P.A.; Chaillé, C. Computer games: Environments for developing spatial cognition? *J. Educ. Comput. Res.* **1987**, *3*, 95–111. [CrossRef]
60. McComas, J.; Pivik, J.; Laflamme, M. Children's transfer of spatial learning from virtual reality to real environments. *CyberPsychol. Behav.* **1998**, *1*, 121–128. [CrossRef]
61. Blank, R.; Barnett, A.L.; Cairney, J.; Green, D.; Kirby, A.; Polatajko, H.; Rosenblum, S.; Smits-Engelsman, B.; Sugden, D.; Wilson, P.; et al. International clinical practice recommendations on the definition, diagnosis, assessment, intervention, and psychosocial aspects of developmental coordination disorder. *Dev. Med. Child Neurol.* **2019**, *61*, 242–285. [CrossRef]
62. World Health Organization. *International Classification of Functioning, Disability and Health: Children and Youth Version: ICF-CY*; World Health Organization: Brussels, Belgium, 2007.
63. Straker, L.M.; Campbell, A.C.; Jensen, L.M.; Metcalf, D.R.; Smith, A.J.; Abbott, R.A.; Pollock, C.M.; Piek, J.P. Rationale, design and methods for a randomised and controlled trial of the impact of virtual reality games on motor competence, physical activity, and mental health in children with developmental coordination disorder. *BMC Public Health* **2011**, *11*, 654. [CrossRef]
64. EbrahimiSani, S.; Sohrabi, M.; Taheri, H.; Agdasi, M.T.; Amiri, S. Effects of virtual reality training intervention on predictive motor control of children with DCD—A randomized controlled trial. *Res. Dev. Disabil.* **2020**, *107*, 103768. [CrossRef]
65. Hoffmann, T.C.; Glasziou, P.P.; Boutron, I.; Milne, R.; Perera, R.; Moher, D.; Altman, D.G.; Barbour, V.; Macdonald, H.; Johnston, M.; et al. Better reporting of interventions: Template for intervention description and replication (TIDieR) checklist and guide. *BMJ* **2014**, *348*, g1687. [CrossRef]

 MDPI

Article

Remote Neuropsychological Intervention for Developmental Dyslexia with the Tachidino Platform: No Reduction in Effectiveness for Older Nor for More Severely Impaired Children

Maria Luisa Lorusso [1,*], Francesca Borasio [1,2] and Massimo Molteni [1]

[1] Unit of Child Psychopathology, Scientific Institute IRCSS E. Medea, 23842 Bosisio Parini, Italy; francesca.borasio@lanostrafamiglia.it (F.B.); massimo.molteni@lanostrafamiglia.it (M.M.)
[2] Department of Psychology, Catholic University of the Sacred Heart, 20123 Milano, Italy
* Correspondence: marialuisa.lorusso@lanostrafamiglia.it

Abstract: Tachidino is a web-platform for remote treatment of reading and writing disorders. A total of 91 children with developmental dyslexia and/or dysorthographia participated in the present study and received Tachidino treatment. The purpose of the study was to compare results obtained after four weeks treatment and a six-month follow-up in older versus younger children and in more versus less severely impaired children (separately subdividing them according to reading speed, reading accuracy, and writing accuracy). The results showed no difference in improvement for reading accuracy and speed in the three age groups, but children below 9 years improved more than older children in writing accuracy. Regarding severity groups, children with more severe initial impairments improved more than children with less severe impairments. Additionally, the results were confirmed after controlling for spurious effects due to use of Z-scores and regression to the mean. The findings are discussed in terms of their theoretical and practical implications.

Keywords: dyslexia; remote intervention; children; age; severity; improvement; follow-up

Citation: Lorusso, M.L.; Borasio, F.; Molteni, M. Remote Neuropsychological Intervention for Developmental Dyslexia with the Tachidino Platform: No Reduction in Effectiveness for Older Nor for More Severely Impaired Children. *Children* **2022**, *9*, 71. https://doi.org/10.3390/children9010071

Academic Editor: Pietro Muratori

Received: 2 December 2021
Accepted: 29 December 2021
Published: 5 January 2022

1. Introduction

Developmental dyslexia is a specific learning disorder affecting reading and spelling that is not due to low intelligence, sensory or neurological damage, or poor educational opportunities [1,2]. Recent research has suggested a multiple-deficit model of dyslexia that could provide a better description of the profiles and risk factors associated with learning difficulties [3,4]. This kind of comprehensive, multi-factor model should be considered to provide an individually tailored treatment program, focused on children's multiple needs [5,6].

Tachidino is a fully automated training programme for reading and writing disorders [7]. It is a web-based platform including a training software, along with systems managing clients' and professionals' data and interfaces. The training software incorporates multi-componential principles, specifically Visual-Attentional Training [8,9] and Visual Hemisphere-Specific Stimulation (VHSS) according to Bakker's Balance Model [10,11] revisited for adaptation to current Italian clinical practice [12]. VHSS stimulates selectively one visual hemisphere to improve reading: this treatment, in isolation, has already been demonstrated to improve both reading fluency and accuracy in dyslexic children [11–17]. The Visual-Attention training is inspired by Action Video Games (AVG), which are characterized by an emphasis on peripheral processing and global perception of stimuli moving at high speed and that are spatio-temporally unpredictable [18,19]. Some studies on children with dyslexia showed that AVGs improve visual-attention abilities, spatial cognition, auditory spatial attention, response time speed, word recognition and phonological decoding efficiency [8,18,20,21].

Tachidino combines the two types of training: in the present training programme, the child is first required to identify and select a moving object among other similar moving objects (visual attention training component) and then to decode or encode words or short text sequences (hemisphere-specific decoding strategies). The text may consist of visually, tachistoscopically presented words/nonwords/combinations or by the same verbal stimuli, auditorily presented through text-to-speech-software. All parameters, such as the laterality of the stimulus, the type of exercise and the list of stimuli, are adapted to the child's characteristics [7]. Auditory presentation was especially added to the program in order to stimulate writing abilities, requiring phoneme-to-grapheme conversion. Indeed, it has been shown [12] that rapid, central rather than lateralized stimulation was the most effective type of stimulation for the improvement of writing abilities, suggesting that writing processes (different from reading) profit more from bilateral stimulation and thus possibly from inter-hemispheric integration.

The remote use of Tachidino allows the maximization of effectiveness in improving reading skills by optimizing the duration and flexibility of the intervention, making the program available to a large group of patients without increasing the costs to health services. Remote treatments for developmental dyslexia can improve reading speed and accuracy after only a few weeks and may foster automatization of the reading process [22]. Moreover, intervention delivered via telepractice appears feasible and engaging, and its effectiveness seems to be comparable to face-to-face programs, although stronger evidence is still needed [23].

There is a general agreement that early identification and intervention are crucial to ensure that children with developmental dyslexia can maximise their educational potential and ameliorate deficits in reading skills [24,25]. However, most studies focused on reading problems in younger children and less is known about training for older children [6]. An Italian study [26] explored whether there may be differences in gains on reading accuracy and fluency between younger and older children with dyslexia (attending primary and secondary school: young group—from the end of the second grade to the end of the fourth grade, versus old group—from the sixth to the eighth grade). These authors found no differences in improvements of reading fluency and accuracy between the two age groups. Moreover, they observed equivalent changes in younger and older children, even if at pretreatment younger children were less accurate and fluent with respect to older children. The demonstration that reading skills may be improved at relatively older ages is an encouraging perspective for children receiving a late diagnosis or treatment. Further studies showed that older children could obtain gains comparable to those of younger students: there is no evidence that older children cannot benefit from specific training [27]. The authors of a recent study evaluated the impact of causes, correlates, and consequences of dyslexia in predicting the outcome of intervention [28]. Among potential impact factors was the severity of reading problems. It was found that children with higher pretreatment word reading skills demonstrated greater improvements compared with peers with more severely impaired reading abilities. These results were in line with those of previous studies e.g., [29].

Based on these premises, it is commonly believed that younger and/or less impaired children are likely to benefit more from intervention on their reading or writing/spelling difficulties. Nonetheless, the results of some studies showing large, clinically relevant improvement in older children, adults, and more severely impaired individuals [26,30–32] in addition to clinical experience led us to reconsider such beliefs.

The purpose of the present study was thus to compare the gains in reading speed and accuracy, and in writing accuracy of different groups of children with a diagnosis of developmental dyslexia. To assess possible age-related differences on the outcomes of the treatment, children were subdivided into three different age groups of similar size, representing different stages of reading acquisition: initial, intermediate, and consolidated. Moreover, to assess the impact of the severity of the impairments on the gains, children were divided into three groups characterized by different levels of impairment in reading and writing. The gains obtained with the treatment for dyslexia were compared both after one month of Tachidino treatment and at follow-up six months after discontinuation of the treatment. Based on clinical experience and on the (few) studies described above, we expected that all groups, irrespective of age and impairment severity, would significantly benefit from the treatment and that no differences would emerge between groups.

2. Materials and Methods

2.1. Participants

A total of 91 children (54 male) aged between 7 and 14 years (mean age = 9.44 years, SD = 1.41) were recruited for this study. Participants had to fulfil the following inclusion criteria: (a) having been diagnosed with Specific Reading Disorder (ICD-10 codes: F81.0 or F81.3) on the basis of standard inclusion and exclusion criteria (ICD-10, WHO, 2011) (at least one Z-score concerning reading/writing speed and/or accuracy below −2); (b) absence of comorbidity with other psychopathological conditions (whereas comorbidity with other learning disorders and/or ADHD was allowed); (c) not having been involved in other clinical intervention programs for learning disorders in the past year. Since the children had been diagnosed, on average, two to three months before the start of the treatment, persistence of reading/writing impairments at pre-test assessment was considered as a confirmation of the stability of the disorder prior to intervention.

Children were selected among patients referred to the IRCCS "E. Medea" and the recruitment took place between January 2018 and January 2020. The present study analyzes data from an observational study for the systematic description of treatment results in a whole, unselected cohort of children with reading and writing disorders in charge at the Institute. The study was approved by the Local Ethics Committee according to the declaration of Helsinki. All parents were informed of the study goals and procedures and their written informed consent was obtained before the beginning of the study. Eighty children (46 male), aged between 7 and 14 (mean age = 9.48 years, SD = 1.44), participated in Time 1, Time 2, and Time 3 assessments. Eleven (12%) participants were no longer available at the time of follow-up, Time 3, due to family reasons not related to the treatment (moving, previous contacts no longer active etc.); performance at pre-test and improvement expressed as D-scores were compared in the two groups, children present and absent at follow-up, and showed no significant differences, all p values > 0.15. A sensitivity analysis (G-power) showed that a sample of 87 participants allows detection of an effect size of 0.17 in a three group, one-tailed, repeated measures ANOVA, within-between interaction, with a power of 0.80. Based on previous results obtained with VHSS [12,13], this effect size was considered acceptable.

The data collection at Time 1, Time 2, and Time 3 of the Tachidino treatment is registered in ClinicalTrials.gov (Code NCT04382482) as an observational study.

2.2. Neuropsychological Tests

The tests commonly employed in the assessment of reading disorders in Italy were used. The results of the tests are expressed as Z-scores according to age norms. The following tests were administered to children in the pre- and post-test and follow-up sessions.

1. Text reading: "Prove di rapidità e correttezza nella lettura del gruppo MT" ("Test of speed and accuracy in reading, developed by the MT group") [33]. This test assesses reading abilities for meaningful texts. It provides separate scores for speed and accuracy. Texts increase in complexity with grade level, and norms are provided for each text. Validity and reliability for the MT text reading test are reported to be satisfactory without further specifications [33]. For the last published version of the test, all test-retest reliability coefficients are above 0.59 and inter-rater reliability coefficients are 1.0 for speed and 0.99 for accuracy [34].
2. Single word/non-word reading: "DDE-2: Batteria per la Valutazione della Dislessia e Disortografia Evolutiva-2" (Assessment battery for Developmental Reading and

Spelling Disorders-2) [35]. The battery assesses speed and accuracy (number of errors) in reading word lists (4 lists of 24 words) and non-word lists (3 lists of 16 non-words), and provides grade norms from the second to the last grade of junior high school. The test has acceptable reliability (mean test-retest coefficients are 0.77 for speed and 0.56 for accuracy).

3. Single word/non-word writing: two writing-to-dictation tasks were taken from the DDE-2 battery, giving accuracy scores according to age norms in writing 48 words and 24 non-words. Although these tests are commonly employed for diagnosis of spelling disorders, no specific reliability and validity data are available.

2.3. *Procedure*

A test-training-retest follow-up experimental design was carried out. All children were tested individually at three time points: Time 1, and Time 2 immediately before and after treatment, while Time 3 occurred 6 months after discontinuation of the treatment. The children's reading and writing skills were assessed by a professional psychologist specialized in cognitive assessment; the psychologist in charge of the assessment was different from psychologists in charge of intervention. Children took part in the remote intervention program for an average total time of 14 h (range 12/18) over a maximum of four weeks. The training does not have a fixed working schedule, so as to adapt to the child's rhythms and attention capacity, but the children are encouraged to work at least 4–5 days per week in sessions of 20–30 min, possibly repeated during the same day, and to read about 20–24 lists of words per week for three/four weeks (the average number of total lists was 73). The exact duration depends on the child's reading level and is about 4–6 h per week. The program includes one pre-treatment meeting in order to define the dyslexia subtype, to demonstrate Tachidino use and programming the first activities, and one intermediate phone call to motivate and monitor correct use of the program.

2.4. *Treatment*

The Tachidino intervention program aims at improving reading through improvement of both decoding strategies and visuo-spatial attentional abilities. Indeed, in each trial of training the child is requested to perform both visual-spatial and word recognition tasks. Firstly, in order to receive the stimulus to be read or listened to, the child is required to recognize the target candy (a spiral candy) among various candies (distractors) and press the spacebar at the exact moment the target candy is crossing a circle target (fixation point). This procedure allows for the control of fixation for precise stimulation of the visual hemifields connected to a certain hemisphere, and also constitutes the visual-attentional training component. If the bar is pressed in the target timeframe and in correspondence of the target candy, the word (or nonword, or word combination) to be decoded/encoded is immediately presented (either visually, in the desired position, or auditorily), and the child is asked to either write the word on the keyboard or re-order a sequence containing all the correct graphemes in random order. This part constitutes the hemisphere-specific stimulation component.

All visual stimuli are presented at tachistoscopic speed to a visual hemifield in order to selectively stimulate the contralateral hemisphere, or they may also be flashed in the center of the computer screen, involving both hemispheres simultaneously. The visual hemisphere-specific stimulation is based on a revisit of Bakker's 'Balance model' [10–13]. Each child was classified as a P-, L-, or M- type dyslexic reader based on the persistent over-reliance on specific reading strategies, on reading speed, and on the pattern of reading errors. More precisely, each child could be included in one of the three following subtypes:

- P-type (decoding strategies based on accurate perceptual analysis mainly supported by the right hemisphere-RH, resulting in slow but relatively accurate reading) if reading speed is at least 1 SD below age mean and the proportion of time-consuming errors over total errors is $\geq$60%;

- L-type (anticipation strategies based on linguistic abilities and mainly supported by the left hemisphere-LH, resulting in relatively fast but inaccurate reading) if reading speed is no more than 1 SD below age mean and the proportion of substantive errors over total errors is $\geq$60%;
- M-type (who strive to use both kinds of strategies but do so inefficiently, resulting in both slow and inaccurate reading) in all other cases (when both error types are present in similar proportion and/or when child is both slow and inaccurate in reading).

The tachistoscopic presentation of visual stimuli depended on classification, and selectively stimulated either the RH (additionally requesting RH-specific perceptual analysis using visually complex materials and/or error detection and correction tasks) or the LH (additionally requesting LH-specific linguistic anticipation using linguistically inter-related materials and/or anticipation/completion tasks). M-type dyslexic readers received the stimulation of the RH first and of the LH at a later stage of treatment, following the stages of natural reading acquisition according to the Balance Model. Central stimulation (and/or auditory presentation) was chosen when the target was to improve writing abilities.

Auditory stimuli were presented through Google text-to-speech synthesis, at the desired speed and pitch according to the child's needs. During auditory presentation of the words, the child is encouraged to extract phonological information from the input, operate phoneme-to-grapheme conversion and match the auditory string with the written visual string (to be either written by the child or reconstructed based on given sequences of the correct graphemes randomly ordered). In the case of auditory stimulation, the hemisphere-specific part is represented in the choice of the material (e.g., low frequency, concrete, highly imageable words for RH stimulation vs. high-frequency, semantically interconnected, abstract words for the LH) and tasks (e.g., correcting errors in word spelling based on auditory input vs. writing or completing semantically related words), addressing either precise (de)coding strategies supported by the RH or anticipation strategies supported by the LH. Auditory presentation is to be considered as a secondary aspect in the training, but relevant for children whose main impairments are in spelling/writing more than in reading skills.

The therapist programs and monitors remotely the child's activities, either in real time or at a different moment, and defines the graphic background, the laterality of the stimuli, the type of exercise (read/write, read/correct, listen/write, listen/correct), the lists of stimuli, exposure times, the characteristics of the font (type, size, spacing, color) and of the speech synthesis (speed and pitch) [7].

2.5. Data Analysis

Three global Z-scores were computed, for pre-, post-test, and follow-up:

(1) global reading speed score, the average of speed Z-scores in text, word and nonword reading, (2) global reading accuracy score, the average of accuracy Z-scores in text, word and nonword reading, and (3) global writing accuracy score, the average of accuracy Z-scores in word and nonword writing to dictation. Subsequently, difference scores between the post-test and the pre-test, the follow-up and the post-test, and the follow-up and the pre-test global scores were calculated, expressing training-related changes.

In order to compare the effectiveness of treatment in different age ranges, children were divided into three age groups: Group 1—younger than 9 years (n = 27), Group 2—between 9 and 10 years (n = 42), Group 3—11 years old and older (n = 22). Moreover, performances were compared between children with different impairment severity levels (degree of reading and writing impairment). In particular, children were divided into those with a performance in the target ability (in turn, reading speed, reading accuracy, and writing accuracy) of -1 Z-scores or higher (Group a), between -3 and -1 Z-scores (Group b) and of -3 Z-scores or lower (Group c). Depending on the specific individual profile, the same child could be included in different groups for different parameters (e.g., one child may have been included in the very severely impaired group with respect to reading speed, to the moderately impaired group for reading accuracy, and in the less severely impaired group for writing accuracy -n are reported in Table 1).

Table 1. Mean scores (SD in parentheses) and effect sizes (Cohen's d_z) of the three age groups.

		Group 1 (Younger than 9 Years)		Group 2 (between 9 and 10 Years)		Group 3 (11 Years Old and Older)	
		Mean (SD)	**Cohen's d_z (p)**	**Mean (SD)**	**Cohen's d_z (p)**	**Mean (SD)**	**Cohen's d_z (p)**
Global reading speed	PRE POST	−2.40 (2.96) −1.61 (2.09)	−0.77 (<0.001)	−1.95 (2.01) −1.05 (1.79)	−0.99 (<0.001)	−2.61 (2.34) −1.51 (1.75)	−0.90 (<0.001)
Global reading accuracy	PRE POST	−2.81 (1.97) −1.47 (1.28)	−1.19 (<0.001)	−2.05 (1.80) −1.17 (1.32)	−0.80 (<0.001)	−2.47 (1.92) −1.51 (1.73)	−0.71 (0.003)
Global writing accuracy	PRE POST	−3.68 (2.91) −1.73 (2.38)	−1.46 (<0.001)	−1.53 (2.84) −0.92 (2.24)	−0.40 (0.015)	−1.22 (1.53) −0.70 (1.45)	−0.56 (0.019)

In order to assess the effectiveness of treatment and the possible treatment-by-group interactions, repeated measures ANOVAs, with age group or severity group as between-subjects factor, and treatment (pre-treatment/post-treatment) as within-subjects factor. Sporadic data are missing due to technical problems during data collection and recording.

3. Results

3.1. Treatment Related Changes: Comparison between Pre- and Post-Test Assessment

3.1.1. Comparison between Age Groups

Comparing pre- and post-test performances with repeated measures ANOVAs with treatment as a within-subject factor and age group as a between-subject factor, a significant main effect of treatment on global reading speed, global reading accuracy, and global writing accuracy confirmed treatment effectiveness ($F(1, 88) = 69.95$, $p < 0.001$, $\eta^2_p = 0.443$; $F(1, 88) = 69.12$, $p < 0.001$, $\eta^2_p = 0.440$; $F(1, 85) = 46.58$, $p < 0.001$, $\eta^2_p = 0.354$, respectively). At post hoc analyses, improvements were significant for all measures within each group (all $p < 0.019$).

For global reading speed and global reading accuracy the analyses showed no significant treatment x age interactions (all $p > 0.27$). However, a significant treatment x age interaction emerged ($F(2, 85) = 9.43$, $p < 0.001$, $\eta^2_p = 0.182$) along with a significant age effect ($F(2, 85) = 4.38$, $p = 0.015$, $\eta^2_p = 0.094$) on global writing accuracy. Tukey post-hoc analysis conducted on the age groups revealed a significant difference between Group 1 (younger than 9 years) and Group 2 (between 9 and 10 years) ($p = 0.031$), and Group 1 and Group 3 (11 years old and older) ($p = 0.030$), indicating that the treatment produced significantly greater improvements in the youngest group of children with respect to both other groups.

To avoid possible spurious effects due to the reduction in variance for reading scores at older ages, a repeated-measures ANOVA was conducted using raw scores instead of z-scores. In particular, syllables/second in text reading was used as a parameter expressing reading speed. Text reading speed was found to be significantly higher after treatment ($F(1, 88) = 182.82$, $p < 0.001$, $\eta^2_p = 0.675$). A post-hoc analysis on syllables/second revealed a significant difference between Group 1 (younger than 9 years) and Group 2 (between 9 and 10 years) ($p = 0.034$). Moreover, a significant treatment x age interaction was found ($F(2, 88) = 3.64$, $p = 0.030$, $\eta^2_p = 0.182$) together with a significant effect of age ($F(2, 88) = 8.25$, $p = 0.001$, $\eta^2_p = 0.158$). At post-hoc analysis, a significant difference emerged between Group 1 (younger than 9 years) and Group 2 (between 9 and 10 years) ($p = 0.008$), and Group 1 and Group 3 (11 years old and older) ($p = 0.001$).

Table 1 presents descriptive results at pre- and post-test assessments.

3.1.2. Comparison between Severity Groups

Pre- and post-test performances in global reading speed were compared across groups differing in reading speed impairment severity, global reading accuracy was compared across groups differing in reading accuracy impairment severity, and global writing accuracy was compared across groups differing in writing accuracy impairment severity. A repeated-measures ANOVAs showed significant main effects of treatment for global reading speed, global reading accuracy, and global writing accuracy (F(1, 88) = 97.2, $p < 0.001$, $\eta^2_p = 0.525$; F(1, 88) = 92.81, $p < 0.001$, $\eta^2_p = 0.513$; F(1, 85) = 90.99, $p < 0.001$, $\eta^2_p = 0.517$, respectively). A significant treatment x severity effect was found for global reading speed (F(2, 88) = 10.31, $p < 0.001$, $\eta^2_p = 0.190$), global reading accuracy (F(2, 88) = 15.24, $p < 0.001$, $\eta^2_p = 0.257$), and global writing accuracy (F(2, 85) = 32.09, $p < 0.001$, $\eta^2_p = 0.430$). Additionally, the analysis clearly showed significant main effects of severity for all measures (F(2, 88) = 22.76, $p < 0.001$, $\eta^2_p = 0.341$ for global reading speed; F(2, 88) = 98.80, $p < 0.001$, $\eta^2_p = 0.692$ for global reading accuracy; F(2, 85) = 73.86, $p < 0.001$, $\eta^2_p = 0.635$ for global writing accuracy).

Tukey post-hoc analysis conducted on the severity groups revealed significant differences in performance in the target ability. Considering global reading speed, the analysis showed differences between Group a (less severely impaired, *n* =) and Group b (moderately impaired) ($p < 0.001$), Group a and Group c (severely impaired) ($p < 0.001$), and Group b and Group c ($p < 0.001$). Figure 1 shows performances in reading speed impairment severity groups, controlling for age.

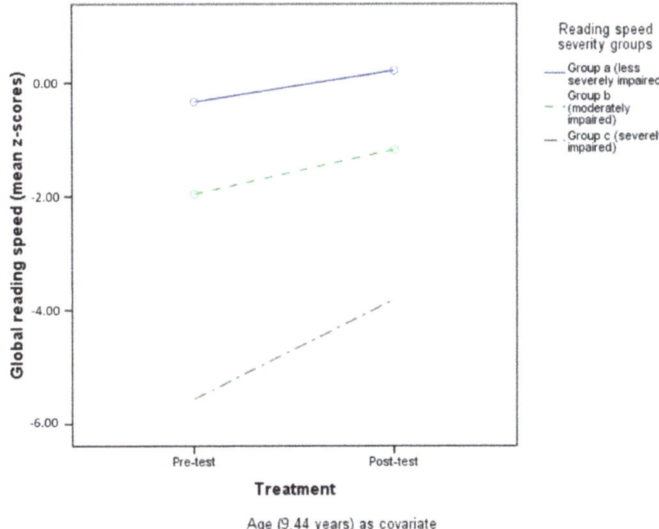

Figure 1. Pre- and post-test reading speed in the three different impairment severity groups.

For global reading accuracy, a difference was found between Group a (less severely impaired) and Group b (moderately impaired) ($p < 0.001$), Group a (less severely impaired) and Group c (severely impaired) ($p < 0.001$), and Group b (moderately impaired) and Group c (severely impaired) ($p < 0.001$). See Figure 2 for performances of reading accuracy impairment severity groups (with age as a covariate).

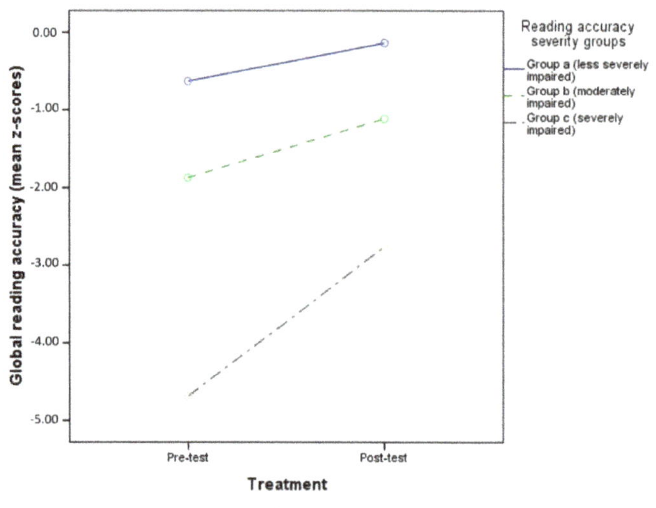

Figure 2. Pre- and post-test reading accuracy in the three different impairment severity groups.

The same differences between impairment severity groups were found for global writing accuracy ($p = 0.001$, $p < 0.001$, $p < 0.001$, respectively for Group a and Group c, Group a and Group b, and Group b and Group c). Figure 3 shows performances in the different writing-accuracy impairment severity groups (age as covariate).

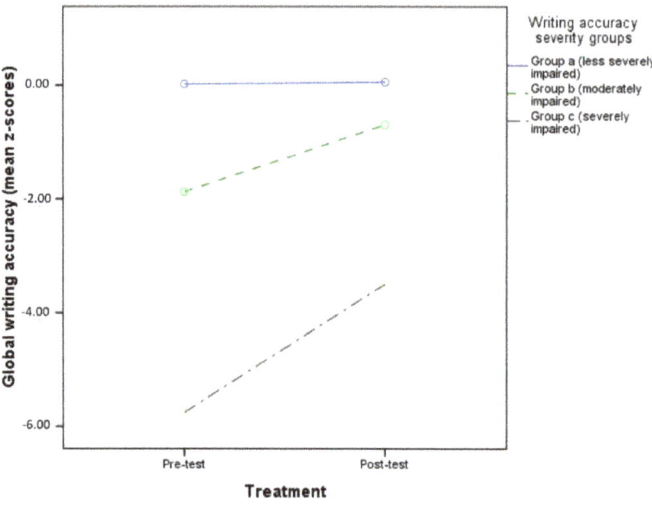

Figure 3. Pre- and post-test writing accuracy in the three different impairment severity groups.

Setting age as a covariate in the same repeated measures ANOVAs did not change the results (treatment x severity: all $p < 0.001$).

In order to avoid spurious effects due to the use of Z-scores, which could result in excessively emphasizing smaller changes in speed or accuracy for older children (for whom the norms show greatly reduced standard deviations with respect to younger children) and to regression to the mean, all comparisons between severity groups yielding significant differences were further checked applying Oldham's (1962) method, see [36]. The results showed that all improvements between the two time points significantly and positively correlated with the average of pre- and post-test measures (all $r_s > 0.430$, all $p < 0.001$). The differences in changes in reading speed/accuracy and in writing accuracy related to starting levels could thus not be due to regression to the mean. Table 2 shows descriptive statistics of the three impairment severity groups.

Table 2. Mean Z-scores (SD in parentheses) and effect sizes (Cohen's d_z) of the three different impairment severity groups (NB for each comparison, the severity of impairment is established based on the parameter being compared: reading speed, reading accuracy, writing accuracy, in turn).

		Group a (Less Severely Impaired)		Group b (Moderately Impaired)		Group c (Severely Impaired)	
		Mean (SD)	Cohen's d_z (p)	Mean (SD)	Cohen's d_z (p)	Mean (SD)	Cohen's d_z (p)
Global reading speed	PRE POST	−0.33 (0.46) 0.20 (0.63)	−0.88 (<0.001) (n = 26)	−1.96 (0.57) −1.17 (0.99)	−0.86 (<0.001) (n = 46)	−5.56 (3.24) −3.81 (2.14)	−1.38 (<0.001) (n = 19)
Global reading accuracy	PRE POST	−0.63 (0.31) −13 (0.48)	−1.06 (<0.001) (n = 24)	−1.87 (0.48) −1.12 (0.75)	−1.06 (<0.001) (n = 40)	−4.69 (1.77) −2.75 (1.53)	−1.20 (<0.001) (n = 27)
Global writing accuracy	PRE POST	0.05 (0.54) 0.02 (0.86)	0.04 (0.83) (n = 37)	−1.87 (0.63) −0.68 (1.01)	−1.30 (<0.001) (n = 28)	−5.79 (2.76) −3.44 (2.76)	−1.42 (<0.001) (n = 23)

3.1.3. Discussion

In the first part of the present study, we compared changes between pre- and post-test in older versus younger children and in more versus less severely impaired children in reading speed/accuracy and writing accuracy. Difference scores between the post-test and the pre-test, expressing training-related changes, were compared between groups of children. Considering the three age groups, no difference in improvement was observed for reading accuracy and speed, while a significant difference emerged for writing accuracy. Precisely, the group of younger children, below 9 years, improved more than the older children in writing accuracy.

The comparison among severity groups showed that, for reading speed, reading accuracy, and writing accuracy, the group of children with more severe initial impairments improved more than groups of children with less severe impairments, and that this was neither due to use of Z-scores producing spurious effects, nor to regression to the mean. A further step was to focus on follow-up performance, in order to evaluate the consolidation of pre-test to post-test gains.

3.2. Consolidation of the Improvements: Comparison of the Pre-Treatment Scores with Follow-Up Scores

3.2.1. Comparison between Age Groups

In ANOVAs comparing pre-treatment and follow-up scores (treatment/consolidation as within factor), and age group as between factor, significant differences were found in global reading speed, global reading accuracy, and global writing accuracy (main effect of consolidation, reflecting the stability of obtained improvements), $F(1, 77) = 64.22.$, $p < 0.001$, $\eta^2_p = 0.455$; $F(1, 77) = 13.85$, $p < 0.001$, $\eta^2_p = 0.152$; $F(1, 76) = 24.40$, $p < 0.001$, $\eta^2_p = 0.243$, respectively). Post-hoc analysis showed significant consolidation effects for all measures in Group 1 (younger than 9 years) (all $p < 0.014$) and in Group 2 (between 9 and 10 years) (all $p < 0.027$). In Group 3 (11 years and older), the stability of improvements was confirmed for global reading speed ($p < 0.001$) and syllables/second in text reading ($p < 0.001$), while global reading accuracy and global writing accuracy were not significant ($p > 0.23$).

For global reading speed and global reading accuracy the analyses showed no significant consolidation x age interactions (all $p > 0.41$). However, a significant consolidation x age interaction emerged for global writing accuracy ($p = 0.006$). No significant age effects emerged (all $p > 0.12$).

Tukey post-hoc analysis was conducted on global writing accuracy, yet showed no significant differences between the three groups (all $p > 0.13$). Additionally, t-tests comparing pre-test to follow-up difference scores in the three groups failed to show any significant difference between pairs of groups (all $p > 0.09$).

3.2.2. Comparison between Severity Groups

The consolidation of improvements from pre-test to follow-up was compared in the three severity groups. A significant consolidation effect emerged for global reading speed ($F(1, 77) = 79.96$, $p < 0.001$, $\eta^2_p = 0.509$), global reading accuracy ($F(1, 77) = 15.33$, $p < 0.001$, $\eta^2_p = 0.166$), and global writing accuracy ($F(1, 76) = 43.87$, $p < 0.001$, $\eta^2_p = 0.366$). Consolidation x severity effects also emerged on the three measures (all $p < 0.019$). Moreover, as expected, a significant severity effect was found for all parameters (all $p < 0.001$).

Tukey post-hoc analysis conducted on the severity groups showed that improvements/consolidation on both global reading speed and global reading accuracy differed between Group a (less severely impaired) and Group b (moderately impaired) ($p < 0.001$), Group a and Group c (severely impaired) ($p < 0.001$), and Group b and Group c ($p < 0.001$). Lastly, global writing accuracy appeared different between Group a (less severely impaired) and Group b (moderately impaired) ($p = 0.001$), Group a and Group c (severely impaired) ($p < 0.001$), and Group b and Group c ($p < 0.001$).

Setting age as a covariate in the same repeated-measures ANOVAs did not change these results (treatment x severity, all $p < 0.001$).

Additional analyses following Oldham's method showed significant correlations between the average of pre-test and follow-up measures and the improvements in global reading speed and global writing accuracy (all $r_s > -0.335$, all $p < 0.003$), while global reading accuracy did not reach significance ($r = -0.145$, $p = 0.201$). In this case, improvements in performances in the three impairment severity groups based on reading accuracy performance could be partially explained by a spurious effect due to regression to the mean.

3.2.3. Discussion

In this second part of the present study, we evaluated the consolidation of the improvements obtained at six months after the end of Tachidino treatment in different groups of children. We observed the stabilization of the improvements obtained in the post-test assessment. Specifically, upon comparison of the three age groups, it was observed that children of the youngest group obtained a greater improvement in writing accuracy with respect to older children.

Considering the differences between impairment severity groups, results showed that, for reading speed, reading accuracy, and writing accuracy, the group of children with more severe initial impairments at pre-test improved more than groups of children with less severe impairments.

3.3. *Treatment Related Changes: Effects of Discontinuation of Treatment between Post-Test and Follow-Up*

3.3.1. Comparison between Age Groups

Comparing post-test and follow-up results, a significant difference on global reading accuracy ($F(1, 77) = 18.19$, $p < 0.001$, $\eta^2_p = 0.191$) emerged, while the other two global measures did not differ between the two time points ($p > 0.23$). In particular, a post-hoc analysis showed differences in global reading accuracy within the three groups (all $p < 0.008$). Neither significant discontinuation x age interactions were found (all $p > 0.32$), nor any age effects (all $p > 0.42$). Figure 4 shows performances at pre-test, post-test, and follow-up of the three age groups.

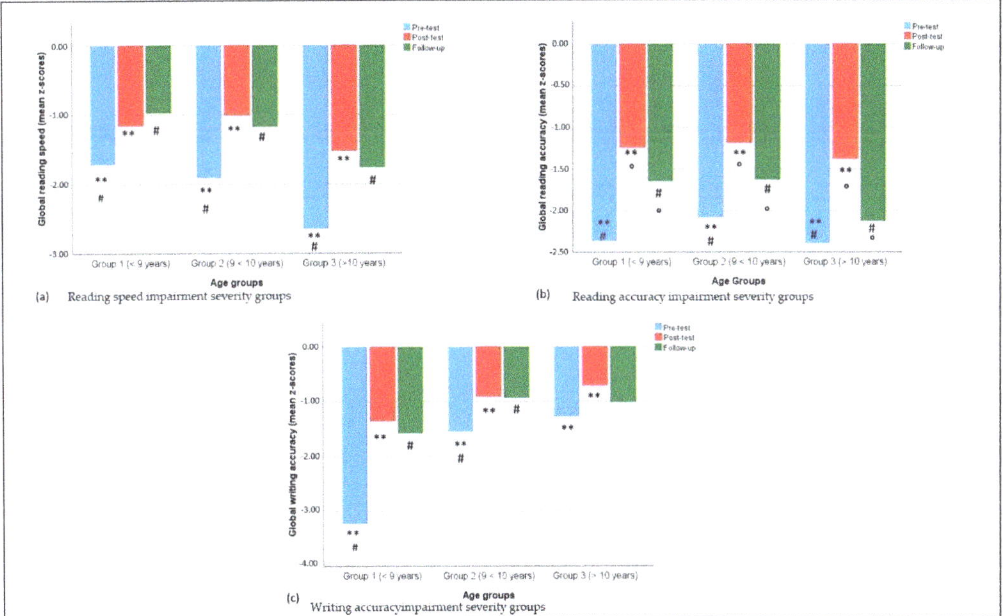

Figure 4. Pre-test, post-test, and follow-up reading and writing scores in the three different age groups. Identical symbols (**, °, #) indicate pairs of significantly different variables within each between-groups comparison.

3.3.2. Comparison between Severity Groups

Regarding possible changes due to discontinuation of treatment between post-test and follow-up, the two performances were compared across impairment severity groups. Global reading speed and global writing accuracy did not differ between post-test and follow-up assessments (all $p > 0.28$), while a significant difference was found for global reading accuracy ($F(1, 77) = 21.68$, $p < 0.001$, $\eta^2_p = 0.220$).

Discontinuation x severity interaction effects were not significant (all $p > 0.06$).

Specific severity effects, as it could be predicted, were found for global reading speed ($F(2, 77) = 68.15$, $p < 0.001$, $\eta^2_p = 0.639$), global reading accuracy ($F(2, 77) = 49.63$, $p < 0.001$, $\eta^2_p = 0.563$), and global writing accuracy ($F(2, 76) = 37.36$, $p < 0.001$, $\eta^2_p = 0.496$). Figure 5 shows performances at pre-test, post-test, and follow-up of the three severity groups.

Additionally, post-hoc analyses showed that the differences in performance level among groups remained significant after treatment (for all pairs, $p < 0.005$). Setting age as a covariate in the same repeated measures ANOVAs did not change the results (treatment x severity all $p > 0.056$).

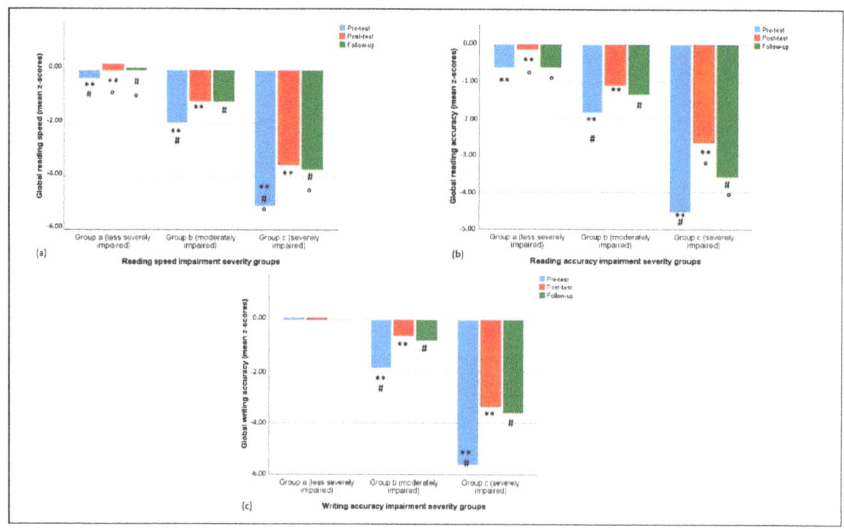

Figure 5. Pre-test, post-test, and follow-up reading and writing scores in the three different impairment severity groups. Identical symbols (**, °, #) indicate pairs of significantly different variables within each between-groups comparison.

3.3.3. Discussion

In the last analyses on possible effects of the discontinuation of treatment on improvements, comparing post-test and follow-up scores, no interaction between discontinuation and severity was found. Global reading accuracy was the only measure that differed between the two time points, both for age groups and impairment severity groups comparisons. We hypothesized that one reason why reading accuracy shows a certain decrease in improvement after post-test is that the reading accuracy score includes text reading, for which different passages are provided for each grade, thus posing greater difficulties to children who had passed to the following grade at follow-up test. In other terms, since the typology and difficulty level of the text reading test increased from one grade to the next grade, children received a different, more complex text at follow-up with respect to pre- and post-test assessments.

To address this hypothesis, a follow-up analysis was performed decomposing the global accuracy score into its component scores. Paired t-tests on Z-scores confirmed that the worsening from post-test to follow-up did not concern word reading accuracy (t (79) = 0.798, $p = 0.427$), while it did concern text reading accuracy (t (78) = 5.475, $p < 0.001$). Nonetheless, decrease in accuracy after post-test also concerned nonword reading (t (78) = 3.624, $p = 0.001$), thus our hypothesis was only partially confirmed. This suggests that reading accuracy is indeed a parameter that suffers more from discontinuation in intervention, and/or longer treatment periods should be foreseen in order to obtain more stable improvements in this respect.

4. General Discussion and Conclusions

The aim of the present study was to compare the beneficial gains in reading and in writing among different groups of children with a diagnosis of developmental dyslexia. Specifically, we assessed possible age-related differences in the outcomes of the treatment, and the impact of the severity of the impairments on the gains. To this aim, participants were subdivided into three different age groups representing different stages of reading acquisition, and three groups characterized by different levels of impairment with respect to reading speed, reading accuracy, and writing accuracy, in turn. All children received a one-month treatment with Tachidino, a web-based program for neuropsychological intervention in developmental dyslexia [7].

In the first part of the study, treatment-related changes were assessed. Considering the three age groups, no difference in improvement was observed for reading accuracy and speed, while a significant difference emerged for writing accuracy. Specifically, the group of younger children, below 9 years of age, improved more than older children in this measure. This was not due to reduction of effectiveness at older age (improvement of writing skills as measured in z-scores were high in all age groups) but rather to a particularly high rate of improvement in the youngest group.

The comparison among impairment severity groups showed that, for all measures, i.e., reading speed, reading accuracy, and writing accuracy, children with more severe initial impairments improved more than groups of children with less severe impairments. This was additionally confirmed after checking for possible spurious effects due to regression to the mean and, for reading speed only (for which raw scores had been recorded as syllables per second), to the use of Z-scores possibly amplifying gains in older children.

In recent years, and mostly during the last months of the COVID-19 pandemic, the use of remote interventions has made it possible to keep rehabilitation programs active, reducing the costs of treatments while increasing the efficiency of reading remediation programs [22,37]. Tachidino is a multi-componential treatment which incorporates two different stimulations: Visual-Attentional Training [8,9] and Visual Hemisphere-Specific Stimulation (VHSS) [10,11]. The effectiveness of the two component trainings had already been shown [9,12,13,16–18] as had the effectiveness of an intensive combined treatment [38]. The present study, using a remote multi-componential treatment where both components are not only added to each other but merged into a single treatment, confirms the enhancements described in Cancer and colleagues [38], which adopted an outpatient remediation program.

In the second part of the study, we explored the consolidation of reading and writing improvements six months after the discontinuation of treatment. Moreover, in that case, different groups of children were compared. The results substantially confirmed those obtained in the post-test assessment. Regarding the comparison of the three age groups of children, the younger children obtained, and maintained, a greater improvement in writing accuracy with respect to both groups of older children. The comparison between the severity groups showed that for all measures, i.e., reading speed, reading accuracy, and writing accuracy, the group of children with more severe initial impairments obtained larger gains than groups of children with less severe impairments. Possible reduction in improvements between post-test and follow-up were also investigated in the different groups. Reading accuracy was the only measure for which it was possible to observe a worsening of performance from post-test to follow-up assessment. It should be highlighted, though, that the difference between pre-test and follow-up was still highly significant, indicating that improvement was maintained and still relevant, even if its size was reduced from post-test to follow-up. These results are in line with those of previous studies showing that remote treatment of dyslexia may improve reading speed and accuracy after only a few weeks of treatment, fostering automatization of the reading process [22]. In addition, a recent study on the effects of computerized cognitive training on visual-spatial working memory and reading performance showed that improvements in attention and visual-spatial working memory lasted for a period of 6 months after treatment [39].

The main findings of the present work relate to the absence of any evidence of reduction in treatment effectiveness for children who are either older or severely impaired. By contrast, older children were found to improve more than younger children in reading speed when using raw scores comparisons. This confirms that there are no objective reasons for limiting treatment to younger children, as what often occurs in clinical practice, nor to children with less severe impairments. The idea that younger children are more responsive to treatment is probably derived from older theories about the reduction in brain plasticity after a certain age, and from research on language impairments, where indeed there is evidence of a developmental window (the so-called "critical period") when response to intervention is more likely to occur and to a larger extent [40]. However, it should be considered that language is a largely different function compared to reading and writing abilities. Indeed, language development follows stable, pre-determined trajectories that are linked to neurobiological development, although it additionally reflects the effects of social interaction and, minimally, of direct instruction [41]. Reading and writing, by contrast, are lately (both phylogenetically and ontogenetically) learned abilities substantially related to formal education, therefore their development is not biologically determined and, consequently, can reasonably be more flexible and subjected to adaptation to environmental conditions. On these grounds, it is not surprising that similar improvements can take place even at later stages of development, such as adolescence, as was also observed in adult populations [26,30,31].

Regarding severity, clearly it could be argued that the more impaired an ability is at an initial stage, the more space there is for improvement. This line of reasoning is probably the simplest explanation for our findings about greater improvement in more severely impaired children. Nonetheless, it should be noted that, in spite of this statistical truth, more severely impaired children are often not offered intervention in clinical practice, favoring other forms of support for their education and academic development (e.g., technological aids to support reading and writing etc.). This is of course justified if observing that a very severe impairment makes it difficult to access school contents through written sources and to express oneself in written form. In this perspective, it is perfectly reasonable that these children should be offered more support and more aids in their school education. This reasoning, though, does not imply that intervention targeted to improve their reading and writing skills, in addition to external and technological support, should be useless or irrelevant. Looking at the large improvements that were observed in these children's scores after only one month of treatment, it is evident that such changes are highly relevant in both absolute and relative terms (as shown by raw scores and z-scores), and that these children are much closer to "normal" performance at post-test than they were at pre-test. Even a change from a very severe impairment to a moderate or non-severe impairment could allow these children to access tasks and activities or experiences that were precluded before treatment, and affect their self-perceptions and self-confidence to a large extent, improving their general quality of life [42,43]. Moreover, further research should clarify whether repeated cycles of treatment could bring further, clinically relevant improvement: existing studies indicate that improvement after the first cycle of intervention is reduced as compared to the former, but still significant. What the present results allow us to exclude quite categorically is that the presence of a more severe impairment reduces the effectiveness of treatment, at least if treatment is conducted according to multi-factorial principles and involves multiple functions. It is possible that other studies which suggested a reduced response to intervention in severely impaired children employed different types of interventions that acted on mechanisms which become less responsive and more rigid over time, or that acted on multiple components allowing for compensatory processes to take place based on different, distributed networks of functions [5,6].

Clearly, the present study does not provide answers to all the questions that were raised, and more research is needed to clarify, for instance, the effects of longer treatments for very severe disorders, or the persistence of improvements after longer periods of discontinuation of treatment. Moreover, some theoretical questions remain open. First of all, the effectiveness observed from the Tachidino program may depend on both components of the training and it is not possible to disentangle the two contributions. We assumed that each of the two components of the treatment produced effects that are similar to those observed in previous studies when each component was applied separately, i.e., relatively larger effects on accuracy through the VHSS [12,13,15,17] and larger effects on speed through the AVG [9,16,18,20] component. The precise, specific contribution of auditory stimulation would need to be investigated through controlled experiments. A comparison of the results of the intervention program with other treatment programs is presented elsewhere

(Lorusso et al., submitted). The absence of a direct comparison of the improvements with those obtained in a control group in the present study constitutes a limitation, nonetheless the stability of improvements at follow-up and in spite of reading different texts (as shown by the comparison of raw scores in text reading tests) suggests that improvement itself can hardly be the result of simple test repetition effects.

Altogether, we believe that the data presented provide a very clear-cut picture of differential, age-related and severity-related constraints on reading and writing improvements, and that these should be considered in clinical practice planning and investments. Even if the present results would require replication and the effectiveness of the Tachidino program has to be demonstrated in comparison with other interventions, the present evidence is highly encouraging with respect to the possibility of a large-scale, even remotely delivered intervention for these highly frequent disorders (3–5% in the Italian school population [44].

Author Contributions: Conceptualization, M.L.L.; methodology, M.L.L.; validation, M.L.L.; formal analysis, M.L.L., F.B.; investigation, M.L.L.; resources, M.M.; data curation, M.L.L., F.B.; writing—original draft preparation, M.L.L., F.B.; writing—review and editing, M.L.L., F.B.; visualization, F.B.; supervision, M.M.; project administration, M.M.; funding acquisition, M.M. All authors have read and agreed to the published version of the manuscript.

Funding: The study was supported by the Italian Ministry of health, grants RC 2019, RC 2020, and RC 2021.

Institutional Review Board Statement: The study was approved by the Ethical Committee of the Scientific Institute IRCCS E. Medea, in accordance with the European Union's Standards of Good Clinical Practice, and in accordance with the Declaration of Helsinki, on 12 December 2019 (prot. N. 76/19-CE).

Informed Consent Statement: Informed consent was obtained from all subjects involved in the study. Written informed consent was obtained from the patients to publish this paper.

Data Availability Statement: Due to privacy issues and as requested by the Ethics Committee, data will be made accessible upon request to the first author and under appropriate agreement conditions.

Acknowledgments: The authors wish to thank Paola Mistò and Laura Pigazzini for their invaluable contribution to treatment supervision and data collection.

Conflicts of Interest: The authors declare no conflict of interest.

References

1. Lyon, G.R.; Shaywitz, S.E.; Shaywitz, B.A. A definition of dyslexia. *Ann. Dyslexia* **2003**, *53*, 1–14. [CrossRef]
2. World Health Organization. *International Statistical Classification of Diseases and Related Health Problems*, 10th ed.; World Health Organization: Geneva, Switzerland, 2010.
3. Ring, J.; Black, J.L. The multiple deficit model of dyslexia: What does it mean for identification and intervention? *Ann. Dyslexia* **2018**, *68*, 104–125. [CrossRef]
4. Pennington, B.F. From single to multiple deficit models of developmental disorders. *Cognition* **2006**, *101*, 385–413. [CrossRef]
5. Pacheco, A.; Reis, A.; Araújo, S.; Inácio, F.; Petersson, K.M.; Faísca, L. Dyslexia heterogeneity: Cognitive profiling of Portuguese children with dyslexia. *Read. Writ.* **2014**, *27*, 1529–1545. [CrossRef]
6. Peterson, R.L.; Pennington, B.F. Developmental dyslexia. *Annu. Rev. Clin. Psychol.* **2015**, *11*, 283–307. [CrossRef] [PubMed]
7. Lorusso, M.L.; Fumagalli, V. La piattaforma Tachidino per la riabilitazione in remoto dei disturbi della lettura e della scrittura [The Tachidino platform for remote rehabilitation of reading and spelling disorders]. In *Teleriabilitazione Nei Disturbi Di Apprendimento*; Bachmann, C., Gagliardi, C., Marotta, L., Eds.; Erickson: Trento, Italy, 2020; pp. 153–170.
8. Bediou, B.; Adams, D.M.; Mayer, R.E.; Tipton, E.; Green, C.S.; Bavelier, D. Meta-analysis of action video game impact on perceptual, attentional, and cognitive skills. *Psychol. Bull.* **2018**, *144*, 77–110. [CrossRef] [PubMed]
9. Franceschini, S.; Gori, S.; Ruffino, M.; Viola, S.; Molteni, M.; Facoetti, A. Action video games make dyslexic children read better. *Curr. Biol.* **2013**, *23*, 462–466. [CrossRef]
10. Bakker, D.J. Neuropsychological classification and treatment of dyslexia. *J. Learn. Disabil.* **1992**, *25*, 102–109. [CrossRef]
11. Bakker, D.J. Treatment of developmental dyslexia: A review. *Pediatr. Rehabil.* **2006**, *9*, 3–13. [CrossRef]
12. Lorusso, M.L.; Facoetti, A.; Bakker, D.J. Neuropsychological treatment of dyslexia: Does type of treatment matter? *J. Learn. Disabil.* **2011**, *44*, 136–149. [CrossRef] [PubMed]
13. Lorusso, M.L.; Facoetti, A.; Paganoni, P.; Pezzani, M.; Molteni, M. Effects of visual hemisphere-specific stimulation versus reading-focused training in dyslexic children. *Neuropsychol. Rehabil.* **2006**, *16*, 194–212. [CrossRef] [PubMed]

14. Koen, B.J. The Neurobiological Development of Reading Fluency. In *Neurodevelopment and Neurodevelopmental Disorder*; Fitzgerald, M., Ed.; IntechOpen: London, UK, 2019; Available online: https://www.intechopen.com/chapters/65561 (accessed on 27 December 2021). [CrossRef]
15. Koen, B.J.; Hawkins, J.; Zhu, X.; Jansen, B.; Fan, W.; Johnson, S. The location and effects of visual hemisphere-specific stimulation on reading fluency in children with the characteristics of dyslexia. *J. Learn. Disabil.* **2018**, *51*, 399–415. [CrossRef] [PubMed]
16. Peters, J.L.; De Losa, L.; Bavin, E.L.; Crewther, S.G. Efficacy of dynamic visuo-attentional interventions for reading in dyslexic and neurotypical children: A systematic review. *Neurosci. Biobehav. Rev.* **2019**, *100*, 58–76. [CrossRef] [PubMed]
17. Tressoldi, P.E.; Vio, C.; Lorusso, M.L.; Facoetti, A.; Iozzino, R. Confronto di efficacia ed efficienza tra trattamenti per il miglioramento della lettura in soggetti dislessici. [Comparison of efficacy and efficiency between treatments for reading improvements in dyslexic subjects]. *Psicol. Clin. Dello Svilupp.* **2003**, *7*, 481–494. [CrossRef]
18. Franceschini, S.; Trevisan, P.; Ronconi, L.; Bertoni, S.; Colmar, S.; Double, K.; Facoetti, A.; Gori, S. Action video games improve reading abilities and visual-to-auditory attentional shifting in English-speaking children with dyslexia. *Sci. Rep.* **2017**, *7*, 5863. [CrossRef] [PubMed]
19. Green, C.S.; Bavelier, D. Action video game modifies visual selective attention. *Nature* **2003**, *423*, 534–537. [CrossRef]
20. Bertoni, S.; Franceschini, S.; Puccio, G.; Mancarella, M.; Gori, S.; Facoetti, A. Action Video Games Enhance Attentional Control and Phonological Decoding in Children with Developmental Dyslexia. *Brain Sci.* **2021**, *11*, 171. [CrossRef]
21. Chopin, A.; Bediou, B.; Bavelier, D. Altering perception: The case of action video gaming. *Curr. Opin. Psychol.* **2019**, *29*, 168–173. [CrossRef]
22. Pecini, C.; Spoglianti, S.; Bonetti, S.; Di Lieto, M.C.; Guaran, F.; Martinelli, A.; Gasperini, F.; Cristofani, P.; Casalini, C.; Mazzotti, S.; et al. Training RAN or reading? A telerehabilitation study on developmental dyslexia. *Dyslexia* **2019**, *25*, 318–331. [CrossRef] [PubMed]
23. Furlong, L.; Serry, T.; Bridgman, K.; Erickson, S. An evidence-based synthesis of instructional reading and spelling procedures using telepractice: A rapid review in the context of COVID-19. *Int. J. Lang. Commun. Disord.* **2021**, *56*, 456–472. [CrossRef]
24. Colenbrander, D.; Ricketts, J.; Breadmore, H.L. Early identification of dyslexia: Understanding the issues. *Lang. Speech Hear. Serv. Sch.* **2018**, *49*, 817–828. [CrossRef]
25. Snowling, M.J. Early identification and interventions for dyslexia: A contemporary view. *J. Res. Spec. Educ. Needs* **2013**, *13*, 7–14. [CrossRef] [PubMed]
26. Tressoldi, P.E.; Lorusso, M.L.; Brenbati, F.; Donini, R. Fluency remediation in dyslexic children: Does age make a difference? *Dyslexia* **2008**, *14*, 142–152. [CrossRef]
27. Lovett, M.W.; Steinbach, K.A. The effectiveness of remedial programs for reading disabled children of different ages: Does the benefit decrease for older children? *Learn. Disabil. Q.* **1997**, *20*, 189–210. [CrossRef]
28. Middleton, A.E.; Farris, E.A.; Ring, J.J.; Odegard, T.N. Predicting and Evaluating Treatment Response: Evidence Toward Protracted Response Patterns for Severely Impacted Students With Dyslexia. *J. Learn. Disabil.* **2021**. [CrossRef]
29. Stuebing, K.K.; Barth, A.E.; Trahan, L.H.; Reddy, R.R.; Miciak, J.; Fletcher, J.M. Are child cognitive characteristics strong predictors of responses to intervention? A meta-analysis. *Rev. Educ. Res.* **2015**, *85*, 395–429. [CrossRef]
30. Eden, G.F.; Jones, K.M.; Cappell, K.; Gareau, L.; Wood, F.B.; Zeffiro, T.A.; Dietz, N.A.; Agnew, J.A.; Flowers, D. Neural changes following remediation in adult developmental dyslexia. *Neuron* **2004**, *44*, 411–422. [CrossRef] [PubMed]
31. Nukari, J.M.; Poutiainen, E.T.; Arkkila, E.P.; Haapanen, M.L.; Lipsanen, J.O.; Laasonen, M.R. Both individual and group-based neuropsychological interventions of dyslexia improve processing speed in young adults: A randomized controlled study. *J. Learn. Disabil.* **2020**, *53*, 213–227. [CrossRef]
32. Suggate, S.P. Why what we teach depends on when: Grade and reading intervention modality moderate effect size. *Dev. Psychol.* **2010**, *46*, 1556–1579. [CrossRef]
33. Cornoldi, C.; Colpo, G. *Nuove Prove Di Lettura MT [New MT Reading Test]*; Giunti O.S: Florence, Italy, 1998.
34. Cornoldi, C.; Carretti, B. *Prove MT-3 Clinica: La Valutazione Delle Abilità Di Lettura E Comprensione [MT-3 Test for Clinical Work: Assessing Reading and Comprehension Abilities]*; Giunti EDU: Florence, Italy, 2016.
35. Sartori, G.; Job, R. *DDE-2: Batteria Per La Valutazione Della Dislessia E Della Disortografia Evolutiva-2 [Assessment Battery for Developmental Reading and Spelling Disorders]*; Giunti O.S: Florence, Italy, 2007.
36. Snider, S.E.; Quisenberry, A.J.; Bickel, W.K. Order in the absence of an effect: Identifying rate-dependent relationships. *Behav. Processes* **2016**, *127*, 18–24. [CrossRef]
37. Spencer-Smith, M.; Klingberg, T. Benefits of a working memory training program for inattention in daily life: A systematic review and meta-analysis. *PLoS ONE* **2015**, *10*, e0119522. [CrossRef]
38. Cancer, A.; Bonacina, S.; Antonietti, A.; Salandi, A.; Molteni, M.; Lorusso, M.L. The effectiveness of interventions for developmental dyslexia: Rhythmic reading training compared with hemisphere-specific stimulation and action video games. *Front. Psychol.* **2020**, *11*, 1158. [CrossRef]
39. Lotfi, S.; Rostami, R.; Shokoohi-Yekta, M.; Ward, R.T.; Motamed-Yeganeh, N.; Mathew, A.S.; Lee, H.J. Effects of computerized cognitive training for children with dyslexia: An ERP study. *J. Neurolinguistics* **2020**, *55*, 100904. [CrossRef]
40. Newport, E.L.; Bavelier, D.; Neville, H.J. Critical thinking about critical periods: Perspectives on a critical period for language acquisition. In *Language, Brain and Cognitive Development: Essays in Honor of Jacques Mehler*; Dupoux, E., Ed.; MIT Press: Cambridge, MA, USA, 2001; pp. 481–502.

41. Bishop, D.V. What causes specific language impairment in children? *Curr. Dir. Psychol. Sci.* **2006**, *15*, 217–221. [CrossRef] [PubMed]
42. Hakkaart-van Roijen, L.; Goettsch, W.G.; Ekkebus, M.; Gerretsen, P.; Stolk, E.A. The cost-effectiveness of an intensive treatment protocol for severe dyslexia in children. *Dyslexia* **2011**, *17*, 256–267. [CrossRef]
43. Glazzard, J. The impact of dyslexia on pupils' self-esteem. *Support Learn.* **2010**, *25*, 63–69. [CrossRef]
44. Barbiero, C.; Montico, M.; Lonciari, I.; Monasta, L.; Penge, R.; Vio, C.; Tressoldi, P.E.; Carrozzi, M.; De Petris, A.; De Cagno, A.G.; et al. The lost children: The underdiagnosis of dyslexia in Italy. A cross-sectional national study. *PLoS ONE* **2009**, *14*, e0210448. [CrossRef]

Article

Dyslexia Telerehabilitation during the COVID-19 Pandemic: Results of a Rhythm-Based Intervention for Reading

Alice Cancer [1,*], Daniela Sarti [2], Marinella De Salvatore [2], Elisa Granocchio [2], Daniela Pia Rosaria Chieffo [3] and Alessandro Antonietti [1]

[1] Department of Psychology, Università Cattolica del Sacro Cuore, 20123 Milan, Italy; alessandro.antonietti@unicatt.it

[2] Developmental Neurology Unit—Language and Learning Disorders Service, Fondazione IRCCS Istituto Neurologico Carlo Besta, 20133 Milan, Italy; daniela.sarti@istituto-besta.it (D.S.); marinella.desalvatore@istituto-besta.it (M.D.S.); elisa.granocchio@istituto-besta.it (E.G.)

[3] Department of Life Sciences and Public Health, Università Cattolica del Sacro Cuore, 00168 Rome, Italy; danielapiarosaria.chieffo@policlinicogemelli.it

* Correspondence: alice.cancer@unicatt.it

Abstract: The COVID-19 outbreak necessitated a reorganization of the rehabilitation practices for Learning Disorders (LDs). During the lockdown phase, telerehabilitation offered the possibility to continue training interventions while enabling social distancing. Given such an advantage of telerehabilitation methods for LDs, clinical research is still needed to test the effectiveness of diverse teletraining approaches by comparing their outcomes with those of face-to-face interventions. To compare the effectiveness of telerehabilitation vs. in-presence rehabilitation of dyslexia, a rhythm-based intervention for reading, called Rhythmic Reading Training (RRT), was tested in a small-scale clinical trial during the lockdown phase of the COVID-19 pandemic. Thirty children aged 8–13 with a diagnosis of developmental dyslexia were assigned to either a telerehabilitation or an in-presence rehabilitation setting and received RRT for 10 biweekly sessions of 45 min, supervised by a trained practitioner. The results showed that both telerehabilitation and in-presence rehabilitation were effective in improving reading and rapid automatized naming in children with dyslexia and that the effects were comparable between settings. Therefore, RRT was found to be effective in spite of the administration method (remote or in-presence). These results confirm the potential of telemedicine for the rehabilitation of LDs. Clinical Trial ID: NCT04995471.

Keywords: dyslexia; telerehabilitation; rhythm; music therapy; intervention

Citation: Cancer, A.; Sarti, D.; De Salvatore, M.; Granocchio, E.; Chieffo, D.P.R.; Antonietti, A. Dyslexia Telerehabilitation during the COVID-19 Pandemic: Results of a Rhythm-Based Intervention for Reading. *Children* **2021**, *8*, 1011. https://doi.org/10.3390/children8111011

Academic Editor: Andrea Martinuzzi

Received: 30 September 2021
Accepted: 3 November 2021
Published: 5 November 2021

1. Introduction

Telemedicine [1] has made it possible to treat patients in their own environment. Since it started and spread, telemedicine has been proven to be a great asset where no easy access to healthcare was possible. A great number of studies have been conducted in the last decade to explore possibilities given by tech tools in adult/older neuropsychological rehabilitation [2–6]. In recent years, technology has been proven to be a great potential resource for developmental age and attention to tech tools in neurodevelopmental disorder rehabilitation has grown fast. Telerehabilitation represents a great advantage when it comes to developmental age. Several studies investigated the effectiveness of brain computer interfaces, virtual reality tools, and computer-based training for Attention Deficit Disorder (ADD) [7,8], Developmental Coordination Disorder (DCD) [9], Autistic Spectrum Disorder (ASD) [10,11], anxiety disorders [12,13], and eating disorders [14,15].

Literature confirms a series of additional advantages given by the use of technology in child rehabilitation. Web-based approaches result, in fact, in increased enthusiasm and a reduced dropout risk. Additionally, a tech approach to rehabilitation could lead to a better generalization of learning (even if no clear agreement is found in literature on this point). Advances in technology are constantly increasing the number of available tools: Bio-neurofeedback, virtual and augmented reality tools, and computer software ensure high motivation in children attending rehabilitation programs because of the playful environment and the quick reward design.

The year 2020 was a worldwide turning point for telemedicine because of the COVID-19 pandemic. Studies on telemedicine have flourished, supported by the urgent need to treat patients while protecting the healthcare of professionals. Telemedicine represents the first-line tool for clinical settings enabling social distancing [16–18]. Today, more than ever, the need for evidence-based healthcare technology is increasing to build a strong and cost-effective telerehabilitation healthcare system [19]. Italy was a first-line trench in facing the pandemic. Children with neurodevelopmental disorders and their families suffered particularly from the lockdown and social isolation that followed. The possibility of continuing rehabilitation treatments online had a central role in preventing psychopathological risks generated by emergency situations in more fragile subjects [20]. Information Computer Technology (ICT) has enabled tailored interventions dedicated to the patient's individual needs through the use of computers, tablets, and other media (e.g., videocalls and messaging apps).

Several telemedicine projects dedicated to psychological healthcare have been implemented in the last year [18,21] and the urgency for telehealth programs has come to light [22–26]. Similar to other psychological interventions, the rehabilitation practices for Learning Disorders (LDs) had to be reorganized in the context of the COVID-19 outbreak. To prevent possible negative outcomes associated with untreated neurodevelopmental disorders, Italian clinical guidelines encourage the early identification and continuous treatments of LDs to reduce risk factors and comorbidity [27].

An Italian example of the reorganization of LD practices imposed by COVID-19 was recently described in an article by Sarti and colleagues [28]. More precisely, the Language and Learning Disorders Service of the Fondazione IRCCS Istituto Neurologico Carlo Besta of Milan, Italy [29] reformulated the children's rehabilitation plan by integrating telemedicine in order to ensure continuity of care for patients. The reorganization involved children with Specific Learning Disorders (SLDs) and preschoolers with Language and Speech Disorders. To select the appropriate patients for telerehabilitation, some feasibility criteria were followed: neuropsychological criteria, attentional, adaptive, and motivational requirements of children as well as network requirements such as connection stability, availability of devices in the family, and the children's and their families' familiarity and autonomy with new technologies. In this regard, the level of competence of children with LDs should not be overestimated, as recent studies have shown that, contrary to expectations, Italian students show little competence in the use of ICT [30].

As for the specific case of reading intervention in developmental dyslexia (DD), several telerehabilitation programs were already available in Italy since before the pandemic. Such programs are based on different developmental reading models. For instance, the online software Reading Trainer [31,32], based on the dual-route developmental reading model [33,34], is a sublexical and lexical treatment aimed to facilitate the correspondences between graphemes and phonemes and to automatize the naming of sublexical units (i.e., syllables and morphemes) and words. Another homebased software is Run the RAN [31,32], a process-oriented intervention addressing difficulties in rapid automatized naming (RAN) tasks, which is one of the main cognitive deficits underlying dyslexia and strongly relates to reading fluency. Run the RAN requires the child to name timed visual nonalphanumeric stimuli (i.e., colors or pictures) as quickly as possible. To remotely improve reading efficiency in children with DD, another group of Italian researchers designed Tachidino [35], a telerehabilitation method based both on hemisphere-specific stimulation, following Bakker's Balance Model [36,37], and on visuospatial selective attention training and the ability to manage visual crowding, according to the magnocellular deficit theory of DD [38,39]. The Tachidino web application includes a large library of Italian words, categorized on the basis of their morpho-linguistic characteristics and specific reading strategies, which are presented tachistoscopically.

The aim of the present study was to compare the effectiveness of telerehabilitation vs. in-presence rehabilitation of reading disorders using a rhythm-based intervention for reading (i.e., Rhythmic Reading Training, RRT).

Extended literature posited the possibility of improving reading abilities in students with DD using music and auditory-based training (for a review, see [40]). Specifically, music activities are supposed to address the phonological difficulties underlying DD through an improved ability to process the temporal components of acoustic stimuli and, ultimately, to improve rhythmic and synchronization abilities. The rhythmic processing of speech cues is indeed fundamental for phonological and reading development in children [41–43], and it is significantly impaired in dyslexia [41,42,44]. A recent literature review by Cancer and Antonietti [40] reported that music-based and auditory-based interventions for dyslexia produced significant effects on phonological abilities, thus supporting the hypothesis of a transfer effect of music training on phonological and reading skills.

Among the reviewed effective interventions, RRT, a rhythm-based training software, was designed to improve reading in Italian students with DD [45–47]. The program includes beat-based reading exercises, in which the reader has to synchronize his/her speech with that of an isochronous beat with an increasing pace. Such an approach was developed to facilitate the segmentation of the phonological units mapped into the metric structure of language by stressing each syllable onset sound and, simultaneously, improving synchronization skills.

The efficacy of RRT was previously tested in several controlled clinical trials, in which the rehabilitation method was delivered to children and adolescents with DD under the supervision of an expert practitioner for 10–20 sessions. More precisely, RRT was found to be effective in improving reading speed and accuracy compared to spontaneous reading development [48]. Furthermore, RRT's efficacy was found to be comparable to that of a traditional reading intervention involving homework [49] and to that of a novel intervention involving visuo-spatial and attentional stimulations [50], with specific larger effects of RRT on pseudo-word reading speed. Finally, significant improvements following RRT were maintained three months after completion of the intervention [46]. Such evidence confirms the feasibility and efficacy of the RRT method for reading training in DD.

Previous applications of RRT were performed in face-to-face settings. To test the possibility of administering RRT remotely, we conducted a small-scale investigation during the lockdown phase of the COVID-19 pandemic. More precisely, we explored the feasibility and effects of RRT remote administration by comparing it with the traditional face-to-face RRT setting.

2. Materials and Methods

2.1. Participants

Children with a specific reading disorder, who had previously received a diagnosis of developmental dyslexia (ICD-10 code: F81.0) [51] based on the LD diagnostic procedure adopted in Italy [52], were recruited from patients of the LD Services of three Italian clinical institutions. Eligibility of participants was determined according to the following inclusion criteria: children between the ages of 8 and 14with a reading performance of > 2 SD below the norm in at least one standardized reading test, normal intelligence (TIQ $\geq$ 80), and the absence of psychiatric and/or neurological conditions.

Thirty children aged between 8 and 13 years (M = 9.89; SD = 1.31; 12 females), who were attending the 2nd to 7th grades, met the inclusion criteria and were included in the study. Parents' written informed consent was obtained prior the recruitment.

2.2. Procedure

The trial protocol was registered on Clinicaltrials.gov (Clinical Trial ID: NCT04995471). Participants were divided into two homogeneous subgroups of the same size (n = 15) using stratified sampling by matching participant for age, sex, TIQ, and reading baseline performance. Each subgroup was then assigned to one of two intervention conditions: telerehabilitation vs. in-presence rehabilitation. Both subgroups received RRT for 10 biweekly sessions of approximately 45 min, supervised by a trained RRT practitioner for a total of 7.5 h of intervention.

To compare the telerehabilitation vs. in-presence methods, a battery of standardized tests, namely, reading and rapid automatized naming tests, were administered to participants before (pre) and immediately after (post) training.

2.3. *Intervention Conditions*

RRT is a computerized training program that includes music-based reading exercises. RRT's activities were designed to address multiple reading subprocesses, specifically syllabic blending, syllabic reading, word recognition, and sublexical decoding [45]. All exercises include a simple rhythmic-melodic stimulation coordinated with a visual cue. The sequential visual selection of each verbal stimuli is synchronized with a regular beat, whose speed is set by the practitioner based on each participant's initial reading level. Speed and difficulty of the exercises are adjusted by the trainer during each session. Practitioners are instructed to gradually increase the speed settings within each activity when the participant reaches a 90% accuracy rate in the previous reading performance. Reading accuracy is assessed by the practitioner during training by counting reading errors in each exercise.

As regards the in-presence administration of RRT, the training sessions took place in a quiet room, where the practitioner and the child were seated next to each other in front of the same computer screen.

In the telerehabilitation setting, RRT was administered using the sharing screen feature of a video teleconferencing platform (i.e., Microsoft Teams). The trainer ran the RRT software on their computer and shared their screen with the participant during a conference call. Prior to the beginning of the intervention, the trainer planned a test call with the participant to check the stability of their internet connection and the quality of the screen sharing procedure.

2.4. *Measures*

Participants' clinical documentation was collected and information about their medical history, diagnosis, and intellectual functioning measures (i.e., Total IQ derived from the Wechsler Intelligence Scale for Children—Fourth Edition [53]) was retrieved from it.

Reading abilities were assessed using standardized tests providing accuracy and speed scores. More precisely, the ability to read aloud age-normed text passages was measured using the 'New MT reading tests for junior high school' [54]. Furthermore, the 'Assessment Battery for Developmental Reading and Spelling Disorders-2' [55] was used to assess word and pseudo-word reading (i.e., 4 lists of 28 words with different lengths and frequency of use; 2 lists of 16 pseudo-words with different lengths).

Finally, rapid automatized naming (RAN) was assessed using the 'Rapid Automatized Naming Test (RAN)—Figures' Test [56], in which participants were required to rapidly and sequentially name a series of black and white figures (i.e., pear, train, dog, star, hand) presented in two 10×5 matrixes. RAN speed (i.e., seconds) was recorded.

2.5. *Statistical Analyses*

A sample size of 30 was calculated to be enough to achieve a power of 0.80 to detect a medium effect size ($\eta^2 = 0.06$) in a mixed factorial ANOVA 2×2—the model which we planned to carry out to test our hypothesis—setting alpha at 0.05.

Assumptions of normality (Kolmogorov–Smirnov ps ranging between 0.25 and 0.99) and homogeneity of error variance (Levene's ps ranging between 0.08 and 0.99) were met for all outcome variables. Therefore, we opted for parametric comparisons (GLMs).

First, we checked that the stratified sampling methods produced homogenous subgroups for age, sex, and baseline reading abilities using the Chi-squared test for categorical variables and independent samples t-tests for continuous variables.

Then, a mixed factorial ANOVA 2 × 2 (Condition: telerehabilitation vs. in-presence; Phase: pre vs. post) was tested for each outcome variable (i.e., reading speed, reading accuracy, RAN). As for the reading measures, composite speed and accuracy scores were computed by averaging text, word, and pseudo-word z-scores.

3. Results

The participants' characteristics are reported in Table 1. Age (t (26) = 0.10; $p = 0.92$), TIQ (t (25) = 0.10; $p = 0.94$), sex ($\chi^2 = 2.22$; $p = 0.14$), and school grade ($\chi^2 = 8.03$; $p = 0.15$) did not differ between groups. In addition, no difference was found between groups in pre-training composite reading measures (reading speed: t (28) = 0.44; $p = 0.66$; reading accuracy: t (28) = 0.19; $p = 0.85$).

Table 1. Participants' characteristics.

	Telerehabilitation	In-Presence Rehabilitation
2nd grade [1]	0	1
3rd grade [1]	2	8
4th grade [1]	5	2
5th grade [1]	4	3
6th grade [1]	3	1
7th grade [1]	1	0
Age [2]	10.30 (1.38)	9.49 (1.16)
Male [1]	7	11
Female [1]	8	4
TIQ [2,3]	98.20 (9.42)	97.85 (15.01)
Baseline reading speed [2,4]	−1.70 (0.63)	−1.80 (0.62)
Baseline reading accuracy [2,4]	−2.89 (2.06)	−3.35 (3.94)

[1] Counts. [2] Mean (Standard Deviations). [3] Total IQ composite score derived from the Wechsler Intelligence Scale for Children—Fourth Edition. [4] Z-scores.

Descriptive statistics of the pre-training and post-training scores for each intervention condition are reported in Table 2.

Table 2. Descriptive statistics of reading and RAN speed pretraining and post-training z-scores, for each rehabilitation condition (i.e., telerehabilitation vs. in-presence rehabilitation).

Condition	Parameter	Phase	M (SD)
Telerehabilitation	Reading speed	Pre	−1.70 (0.63)
		Post	−1.21 (0.72)
	Reading accuracy	Pre	−2.89 (2.06)
		Post	−2.47 (2.42)
	RAN	Pre	−1.82 (1.18)
		Post	−1.19 (1.86)
In-presence rehabilitation	Reading speed	Pre	−1.80 (0.62)
		Post	−1.34 (0.83)
	Reading accuracy	Pre	−3.35 (3.94)
		Post	−2.30 (2.29)
	RAN	Pre	−3.02 (2.69)
		Post	−1.71 (1.14)

Reading speed and reading accuracy improved after training in both conditions, as confirmed by significant Phase main effects (reading speed: $F(1,28) = 70.58$; $p < 0.001$; $\eta^2 = 0.11$; reading accuracy: $F(1,28) = 4.55$; $p < 0.04$; $\eta^2 = 0.02$). Conversely, the Phase × Condition interaction effect was nonsignificant for both reading outcomes (reading speed: $F(1,28) = 0.09$; $p = 0.77$; reading accuracy: $F(1,28) = 0.84$; $p = 0.37$), thus showing no telerehabilitation vs. in-presence rehabilitation difference.

As for the secondary outcome measure, similar results were found for RAN speed, with a significant Phase main effect ($F(1,28) = 5.45$; $p < 0.04$; $\eta^2 = 0.07$) and a nonsignificant Phase × Condition interaction effect ($F(1,28) = 0.67$; $p = 0.43$).

4. Discussion and Conclusions

During the unprecedented events associated with the spread of the COVID-19 disease, telemedicine constituted a chance to treat neurodevelopmental disorders through tailored and goal-oriented interventions, while maintaining social distancing. Given the evident advantages of telerehabilitation methods, clinical research is needed to test their effectiveness in the intervention of LDs, such as DD, by comparing the outcomes with those of traditional face-to-face interventions.

Previous Italian literature analyzed the efficacy of several telerehabilitation methods for DD, showing promising results. In a study that included 34 children with DD attending primary or secondary schools, Tucci and colleagues [57] confirmed the efficacy of an intervention using the online software Reading Trainer [31,32]. After a training period of approximately 13 weeks, with 15 min training sessions at least 3 times a week, the authors found a significant improvement of reading fluency and accuracy. Another study by Pecini and colleagues [32] compared the effects of the Run the RAN training with those of the Reading Trainer, administered in 5–15 min sessions for a total of 8 and 12 h respectively in a group of 45 children with DD. The results showed that the reading speed and accuracy improved regardless of the training type and that improvements lasted for 3 months after the end of the intervention.

Although there were significant within-group effects of telerehabilitation, these studies did not include an in-presence control condition. To our knowledge, the only Italian study which compared telerehabilitation and in-presence rehabilitation of DD is a recently published study by Lorusso and colleagues [35]. The authors compared 65 children with DD who underwent a remote intervention using the Tachidino web application with 49 children who received an in-presence visuospatial intervention (i.e., Action Video Games training [58] in combination with Visual Hemispheric Specific Stimulation [39]). The results showed that improvements emerged in both groups in terms of both speed and accuracy of reading.

The present small-scale clinical trial is the first attempt to measure the specific effect of the online remote administration of an Italian reading training by comparing parallel training conditions with matching materials, activities, procedures, and training schedules. The results of the present study showed that both telerehabilitation and in-presence rehabilitation of reading abilities using a rhythm-based computerized intervention were effective in improving reading and RAN abilities in students with DD and that the effects were comparable between settings. Such results demonstrate that RRT is effective in spite of the administration method (remote or in-presence), thus adding evidence of its potential as a rehabilitation method for DD.

In comparison with other telerehabilitation methods, the RRT makes it possible to achieve significant progress in a shorter period of time, namely, 7.5 h. In addition, the need for support by the parent/caregiver appears to be less significant compared to other methods (e.g., reading errors were recorded by the online trainer during rehabilitation sessions, and no parental supervision was necessary).

Furthermore, these results confirmed the telemedicine potential for the rehabilitation of LDs, which was previously highlighted in several studies conducted during the COVID-19 pandemic. Numerous papers reported on the results obtained in telemedicine satisfaction questionnaires that were completed by adult patients and children's caregivers and that highlighted an overall good level of satisfaction concerning the remodeling of services [29]. Telerehabilitation for LDs has made it possible to maintain therapeutic continuity for younger patients. In particular, the easy remote adaptability of the RRT was a strong point during a period that required rapidity/timeliness in changes within the rehabilitation practices settings [59]. Concerning the reception of the rhythm-based intervention, spontaneous comments by participants revealed that children generally accepted the teletraining positively, showing curiosity towards RRT and expressing their satisfaction with regards to the results obtained. Consistently, parents expressed overall satisfaction with regard to maintaining an improving trend with their children, as anecdotally recorded, given the concerns that families of children with neurodevelopmental disorders had during the pandemic period [60].

As regards telerehabilitation, the remote setting was characterized by both strengths and limitations. On the one hand, the home context where the rehabilitation took place was perceived as reassuring and comfortable for children. On the other hand, the relational components of rehabilitation were less controllable compared to face-to-face interactions. In some cases, the unfamiliarity with new technologies did represent a limit, since children were not autonomous and parents were required to help them. However, this occurred more frequently at the beginning of the remote training path and most children learned quite early to manage the telerehabilitation setting autonomously. We suggest that future investigations should include a preintervention tech training for children who may struggle with the use of technology. Furthermore, for the most fragile patients, such as younger children, children with more severe reading difficulties, and a reduced attention span, the teletraining was more tiring and several interruptions were required during the sessions. As for the other studies presented on the telerehabilitation of DD in regular orthography, the present study confirms that homebased software may foster the automatization of the reading processes after only a few months of teletraining.

Besides the small sample size, one of the limitations of the present study was the lack of follow-up measures, which would assess the long-term effects of the intervention. However, a previous study on RRT showed that reading gains after intervention lasted for three months [46]. Although we expect that such results should be replicated in both in-presence and telerehabilitation settings, future investigations may compare the long-term effects of RRT in different administration conditions.

Finally, despite the positive outcomes of telerehabilitation, as shown by the results of the present study, we believe that reading rehabilitation is better administered in a presence-remote hybrid setting. While telemedicine can facilitate the rehabilitation of LDs in certain conditions, for the most fragile patients, face-to-face treatment is preferable due to the irreplaceable interpersonal variables within the clinical relationship.

Author Contributions: Conceptualization, A.C. and A.A.; Data curation, A.C. and M.D.S.; Formal analysis, A.C.; Investigation, D.S. and M.D.S.; Methodology, A.C., D.S., M.D.S. and E.G.; Project administration, D.S., E.G., D.P.R.C. and A.A.; Supervision, E.G., D.P.R.C. and A.A.; Writing—original draft, A.C., D.S., M.D.S., E.G. and D.P.R.C.; Writing—review and editing, A.C. and A.A. All authors have read and agreed to the published version of the manuscript.

Funding: This research received no external funding.

Institutional Review Board Statement: The study was conducted according to the guidelines of the Declaration of Helsinki and approved by the Psychology Research Ethics Committee of the Università Cattolica del Sacro Cuore, Milan, Italy (Approval Number: 30–18; date of approval: 23 June 2018) and by the Ethics Committee of the Fondazione IRCCS Istituto Neurologico Carlo Besta of Milan, Italy (Approval Number: 82/2021; date of approval: 17 March 2021).

Informed Consent Statement: Informed consent was obtained from all parents of the children involved in the study.

Data Availability Statement: The data presented in this study are available on request from the corresponding author.

Acknowledgments: We thank Serena Germagnoli, Gabriella Pace, Alice Balestri, Giuseppina Corvace, Marco Guida, and Lucia Covarelli who, as specialized trainers and research assistants, participated in the application of the RRT treatment.

Conflicts of Interest: The authors declare no conflict of interest.

References

1. American Telemedicine Association. What Is Telemedicine? The Current State of Virtual Healthcare | Shape. Available online: https://www.shape.com/lifestyle/mind-and-body/what-is-telemedicine (accessed on 13 September 2021).
2. Laver, K.E.; Lange, B.; George, S.; Deutsch, J.E.; Saposnik, G.; Crotty, M. Virtual Reality for Stroke Rehabilitation. *Cochrane Database Syst. Rev.* **2017**. [CrossRef]
3. Corbetta, D.; Imeri, F.; Gatti, R. Rehabilitation That Incorporates Virtual Reality Is More Effective than Standard Rehabilitation for Improving Walking Speed, Balance and Mobility after Stroke: A Systematic Review. *J. Physiother.* **2015**, *61*, 117–124. [CrossRef] [PubMed]
4. Maggio, M.G.; Russo, M.; Cuzzola, M.F.; Destro, M.; La Rosa, G.; Molonia, F.; Bramanti, P.; Lombardo, G.; De Luca, R.; Calabrò, R.S. Virtual Reality in Multiple Sclerosis Rehabilitation: A Review on Cognitive and Motor Outcomes. *J. Clin. Neurosci.* **2019**, *65*, 106–111. [CrossRef]
5. Maresca, G.; Maggio, M.G.; De Luca, R.; Manuli, A.; Tonin, P.; Pignolo, L.; Calabrò, R.S. Tele-Neuro-Rehabilitation in Italy: State of the Art and Future Perspectives. *Front. Neurol.* **2020**, *11*, 563375. [CrossRef] [PubMed]
6. Block, V.A.J.; Pitsch, E.; Tahir, P.; Cree, B.A.C.; Allen, D.D.; Gelfand, J.M. Remote Physical Activity Monitoring in Neurological Disease: A Systematic Review. *PLoS ONE* **2016**, *11*, e0154335. [CrossRef]
7. Spencer, T.; Noyes, E.; Biederman, J. Telemedicine in the Management of ADHD: Literature Review of Telemedicine in ADHD. *J. Atten. Disord.* **2020**, *24*, 3–9. [CrossRef]
8. McGrath, J. ADHD and Covid-19: Current Roadblocks and Future Opportunities. *Ir. J. Psychol. Med.* **2020**, *37*, 204–211. [CrossRef] [PubMed]
9. Lino, F.; Arcangeli, V.; Chieffo, D.P.R. The Virtual Challenge: Virtual Reality Tools for Intervention in Children with Developmental Coordination Disorder. *Children* **2021**, *8*, 270. [CrossRef]
10. Mesa-Gresa, P.; Gil-Gómez, H.; Lozano-Quilis, J.-A.; Gil-Gómez, J.-A. Effectiveness of Virtual Reality for Children and Adolescents with Autism Spectrum Disorder: An Evidence-Based Systematic Review. *Sensors* **2018**, *18*, 2486. [CrossRef] [PubMed]
11. Berenguer, C.; Baixauli, I.; Gómez, S.; Andrés, M.d.E.P.; De Stasio, S. Exploring the Impact of Augmented Reality in Children and Adolescents with Autism Spectrum Disorder: A Systematic Review. *Int. J. Environ. Res. Public Health* **2020**, *17*, 6143. [CrossRef]
12. Vigerland, S.; Lenhard, F.; Bonnert, M.; Lalouni, M.; Hedman, E.; Ahlen, J.; Olén, O.; Serlachius, E.; Ljótsson, B. Internet-Delivered Cognitive Behavior Therapy for Children and Adolescents: A Systematic Review and Meta-Analysis. *Clin. Psychol. Rev.* **2016**, *50*, 1–10. [CrossRef] [PubMed]
13. Hammond, D.C. Neurofeedback with Anxiety and Affective Disorders. *Child Adolesc. Psychiatr. Clin.* **2005**, *14*, 105–123. [CrossRef]
14. Sproch, L.E.; Anderson, K.P. Clinician-Delivered Teletherapy for Eating Disorders. *Psychiatr. Clin. N. Am.* **2019**, *42*, 243–252. [CrossRef]
15. Riva, G. The Key to Unlocking the Virtual Body: Virtual Reality in the Treatment of Obesity and Eating Disorders. *J. Diabetes Sci. Technol.* **2011**, *5*, 283–292. [CrossRef] [PubMed]
16. Bashshur, R.; Doarn, C.R.; Frenk, J.M.; Kvedar, J.C.; Woolliscroft, J.O. Telemedicine and the COVID-19 Pandemic, Lessons for the Future. *Telemed. E-Health* **2020**, *26*, 571–573. [CrossRef]
17. Coffey, J.D.; Christopherson, L.A.; Glasgow, A.E.; Pearson, K.K.; Brown, J.K.; Gathje, S.R.; Sangaralingham, L.R.; Carmona Porquera, E.M.; Virk, A.; Orenstein, R.; et al. Implementation of a Multisite, Interdisciplinary Remote Patient Monitoring Program for Ambulatory Management of Patients with COVID-19. *NPJ Digit. Med.* **2021**, *4*, 1–11. [CrossRef] [PubMed]
18. Provenzi, L.; Grumi, S.; Borgatti, R. Alone With the Kids: Tele-Medicine for Children With Special Healthcare Needs During COVID-19 Emergency. *Front. Psychol.* **2020**, *11*, 1–6. [CrossRef]
19. Galea, M.D. Telemedicine in Rehabilitation. *Phys. Med. Rehabil. Clin.* **2019**, *30*, 473–483. [CrossRef] [PubMed]
20. Masi, A.; Mendoza Diaz, A.; Tully, L.; Azim, S.I.; Woolfenden, S.; Efron, D.; Eapen, V. Impact of the COVID -19 Pandemic on the Well-being of Children with Neurodevelopmental Disabilities and Their Parents. *J. Paediatr. Child Health* **2021**, *57*, 631–636. [CrossRef]
21. Bokolo, A. Jnr. Use of Telemedicine and Virtual Care for Remote Treatment in Response to COVID-19 Pandemic. *J. Med. Syst.* **2020**, *44*, 132. [CrossRef]
22. Summers, J.; Baribeau, D.; Mockford, M.; Goldhopf, L.; Ambrozewicz, P.; Szatmari, P.; Vorstman, J. Supporting Children With Neurodevelopmental Disorders During the COVID-19 Pandemic. *J. Am. Acad. Child Adolesc. Psychiatry* **2021**, *60*, 2–6. [CrossRef]
23. Schachter, S.C. Chapter 21—New technologies and future trends. In *Handbook of Clinical Neurology;* Gallagher, A., Bulteau, C., Cohen, D., Michaud, J.L., Eds.; Neurocognitive Development: Disorders and Disabilities; Elsevier: Amsterdam, The Netherlands, 2020; Volume 174, pp. 289–297.
24. Valentino, L.A.; Skinner, M.W.; Pipe, S. The Role of Telemedicine in the Delivery of Healthcare in the COVID-19 Pandemic. *Haemophilia* **2020**, *26*, e230–e231. [CrossRef]

25. Guido-Estrada, N.; Crawford, J. Embracing Telemedicine: The Silver Lining of a Pandemic. *Pediatr. Neurol.* **2020**, *113*, 13–14. [CrossRef] [PubMed]
26. Bachmann, C.; Gagliardi, C.; Marotta, L. *Teleriabilitazione Nei Disturbi Di Apprendimento*; Erickson: Trento, Itally, 2020; ISBN 978-88-590-2488-0.
27. ISS Consensus Conference "Disturbi Specifici Dell'apprendimento". Available online: https://www.Aiditalia.Org (accessed on 4 August 2021).
28. Sarti, D.; De Salvatore, M.; Gazzola, S.; Pantaleoni, C.; Granocchio, E. So Far so Close: An Insight into Smart Working and Telehealth Reorganization of a Language and Learning Disorders Service in Milan during COVID-19 Pandemic. *Neurol. Sci.* **2020**, *41*, 1659–1662. [CrossRef] [PubMed]
29. Pareyson, D.; Pantaleoni, C.; Eleopra, R.; De Filippis, G.; Moroni, I.; Freri, E.; Zibordi, F.; Bulgheroni, S.; Pagliano, E.; Sarti, D.; et al. Neuro-Telehealth for Fragile Patients in a Tertiary Referral Neurological Institute during the COVID-19 Pandemic in Milan, Lombardy. *Neurol. Sci.* **2021**, *42*, 2637–2644. [CrossRef]
30. Caccia, M.; Giorgetti, M.; Toraldo, A.; Molteni, M.; Sarti, D.; Vernice, M.; Lorusso, M.L. ORCA.IT: A New Web-Based Tool for Assessing Online Reading, Search and Comprehension Abilities in Students Reveals Effects of Gender, School Type and Reading Ability. *Front. Psychol.* **2019**, *10*, 2433. [CrossRef]
31. Pecini, C.; Spoglianti, S.; Michetti, S.; Bonetti, S.; Di, M.L.; Gasperini, F.; Cristofani, P.; Bozza, M.; Brizzolara, D.; Casalini, C.; et al. Telerehabilitation in Developmental Dyslexia: Methods of Implementation and Expected Results. *Minerva Pediatr.* **2018**, *70*, 529–538. [CrossRef]
32. Pecini, C.; Spoglianti, S.; Bonetti, S.; Di Lieto, M.C.; Guaran, F.; Martinelli, A.; Gasperini, F.; Cristofani, P.; Casalini, C.; Mazzotti, S.; et al. Training RAN or Reading? A Telerehabilitation Study on Developmental Dyslexia. *Dyslexia* **2019**. [CrossRef]
33. Coltheart, M.; Curtis, B.; Atkins, P.; Haller, M. Models of Reading Aloud: Dual-Route and Parallel-Distributed-Processing Approaches. *Psychol. Rev.* **1993**, *100*, 589–608. [CrossRef]
34. Frith, U. A Developmental Framework for Developmental Dyslexia. *Ann. Dyslexia* **1986**, *36*, 67–81. [CrossRef]
35. Lorusso, M.L.; Fumagalli, V. La piattaforma "Tachidino" per la riabilitazione in remoto dei disturbi della lettura e della scrittura. In *Teleriabilitazione nei Disturbi di Apprendimento*; Bachmann, C., Gagliardi, C., Marotta, L., Eds.; Erickson: Trento, Italy, 2020.
36. Bakker, D.J. Neuropsychological Classification and Treatment of Dyslexia. *J. Learn. Disabil.* **1992**, *25*, 102–109. [CrossRef]
37. Lorusso, M.L.; Facoetti, A.; Bakker, D.J. Neuropsychological Treatment of Dyslexia: Does Type of Treatment Matter? *J. Learn. Disabil.* **2011**, *44*, 136–149. [CrossRef]
38. Lorusso, M.L.; Facoetti, A.; Molteni, M. Hemispheric, Attentional, and Processing Speed Factors in the Treatment of Developmental Dyslexia. *Brain Cogn.* **2004**, *55*, 341–348. [CrossRef]
39. Lorusso, M.L.; Facoetti, A.; Paganoni, P.; Pezzani, M.; Molteni, M. Effects of Visual Hemisphere-Specific Stimulation versus Reading-Focused Training in Dyslexic Children. *Neuropsychol. Rehabil.* **2006**, *16*, 194–212. [CrossRef]
40. Cancer, A.; Antonietti, A. Music-Based and Auditory-Based Interventions for Reading Difficulties: A Literature Review. *Heliyon Psychol.*. under revision.
41. Huss, M.; Verney, J.P.; Fosker, T.; Mead, N.; Goswami, U. Music, Rhythm, Rise Time Perception and Developmental Dyslexia: Perception of Musical Meter Predicts Reading and Phonology. *Cortex* **2011**, *47*, 674–689. [CrossRef]
42. Flaugnacco, E.; Lopez, L.; Terribili, C.; Zoia, S.; Buda, S.; Tilli, S.; Monasta, L.; Montico, M.; Sila, A.; Ronfani, L.; et al. Rhythm Perception and Production Predict Reading Abilities in Developmental Dyslexia. *Front. Hum. Neurosci.* **2014**, *8*, 1–14. [CrossRef]
43. Ozernov-Palchik, O.; Wolf, M.; Patel, A.D. Relationships between Early Literacy and Nonlinguistic Rhythmic Processes in Kindergarteners. *J. Exp. Child Psychol.* **2018**, *167*, 354–368. [CrossRef] [PubMed]
44. Goswami, U.; Huss, M.; Mead, N.; Fosker, T.; Verney, J.P. Perception of Patterns of Musical Beat Distribution in Phonological Developmental Dyslexia: Significant Longitudinal Relations with Word Reading and Reading Comprehension. *Cortex* **2013**, *49*, 1363–1376. [CrossRef] [PubMed]
45. Cancer, A.; Bonacina, S.; Lorusso, M.L.; Lanzi, P.L.; Antonietti, A. Rhythmic Reading Training (RRT): A computer-assisted intervention program for dyslexia. In *Pervasive Computing Paradigms for Mental Health*; Serino, S., Matic, A., Giakoumis, D., Lopez, G., Cipresso, P., Eds.; Communications in Computer and Information Science; Springer International Publishing: New York, NY, USA, 2016; pp. 249–258. ISBN 978-3-319-32269-8.
46. Cancer; Stievano; Pace; Colombo. Antonietti Cognitive Processes Underlying Reading Improvement during a Rhythm-Based Intervention. A Small-Scale Investigation of Italian Children with Dyslexia. *Children* **2019**, *6*, 91. [CrossRef] [PubMed]
47. Cancer, A.; Antonietti, A. Remedial Interventions for Developmental Dyslexia: How Neuropsychological Evidence Can Inspire and Support a Rehabilitation Training. *Neuropsychol. Trends* **2017**, *22*, 73–95. [CrossRef]
48. Bonacina, S.; Cancer, A.; Lanzi, P.L.; Lorusso, M.L.; Antonietti, A. Improving Reading Skills in Students with Dyslexia: The Efficacy of a Sublexical Training with Rhythmic Background. *Front. Psychol.* **2015**, *6*, 1–8. [CrossRef] [PubMed]
49. Cancer, A.; Stievano, G.; Pace, G.; Colombo, A.; Brembati, F.; Donini, R.; Antonietti, A. Remedial Interventions for Italian Children with Developmental Dyslexia: Comparing the Rhythmic Reading Training to the 'Abilmente' Approach. *Psychol. Music.* under revision.
50. Cancer, A.; Bonacina, S.; Antonietti, A.; Salandi, A.; Molteni, M.; Lorusso, M.L. The Effectiveness of Interventions for Developmental Dyslexia: Rhythmic Reading Training Compared with Hemisphere-Specific Stimulation and Action Video Games. *Front. Psychol.* **2020**, *11*. [CrossRef] [PubMed]

51. *World Health Organization The ICD-10 Classification of Mental and Behavioural Disorders: Clinical Descriptions and Diagnostic Guidelines*; World Health Organization: Geneva, Switzerland, 1992; ISBN 978-92-4-154422-1.
52. PARCC Documento d'intesa: Raccomandazioni Cliniche Sui DSA. Available online: http://www.lineeguidadsa.it (accessed on 4 August 2021).
53. Weschler, D. *Wechsler Intelligence Scale for Children-Fourth Edition (WISC-IV) Administration and Scoring Manual*; The Psychological Association: San Antonio, TX, USA, 2003.
54. Cornoldi, C.; Colpo, G. *Nuove Prove di Lettura MT per la Scuola Media Inferiore: Manuale [New MT Reading Test for Junior High School: Manual]*; Giunti O.S.: Florence, Italy, 1995; ISBN 978-88-09-40085-6.
55. Sartori, G.; Job, R. *DDE-2: Batteria per la Valutazione della Dislessia e della Disortografia Evolutiva-2 [Assessment Battery for Developmental Reading and Spelling Disorders]*; Giunti O.S.: Florence, Italy, 2007; ISBN 978-88-09-40301-7.
56. De Luca, M.; Di Filippo, G.; Judica, A.; Spinelli, D.; Zoccolotti, P. *Test Di Denominazione Rapida e Ricerca Visiva Di Colori, Figure e Numeri. [Rapid Naming Test and Visual Search of Colours, Figures, and Numbers]*; IRCCS Fondazione Santa Lucia: Rome, Italy, 2005.
57. Tucci, R.; Savoia, V.; Bertolo, L.; Vio, C.; Tressoldi, P.E. Efficacy and Efficiency Outcomes of a Training to Ameliorate Developmental Dyslexia Using the Online Software Reading Trainer®. *BPA-Appl. Psychol. Bull. (Boll. Di Psicol. Appl.)* **2015**, *63*, 53–60.
58. Franceschini, S.; Gori, S.; Ruffino, M.; Viola, S.; Molteni, M.; Facoetti, A. Action Video Games Make Dyslexic Children Read Better. *Curr. Biol.* **2013**, *23*, 462–466. [CrossRef]
59. Prvu Bettger, J.; Thoumi, A.; Marquevich, V.; De Groote, W.; Rizzo Battistella, L.; Imamura, M.; Delgado Ramos, V.; Wang, N.; Dreinhoefer, K.E.; Mangar, A.; et al. COVID-19: Maintaining Essential Rehabilitation Services across the Care Continuum. *BMJ Glob. Health* **2020**, *5*, e002670. [CrossRef]
60. Asbury, K.; Fox, L.; Deniz, E.; Code, A.; Toseeb, U. How Is COVID-19 Affecting the Mental Health of Children with Special Educational Needs and Disabilities and Their Families? *J. Autism Dev. Disord.* **2021**, *51*, 1772–1780. [CrossRef]

Article

Telerehabilitation and Wellbeing Experience in Children with Special Needs during the COVID-19 Pandemic

Daniela Sarti [1], Marinella De Salvatore [1], Emanuela Pagliano [1], Elisa Granocchio [1], Daniela Traficante [2] and Elisabetta Lombardi [2,*]

1 Foundation IRCCS Neurological Institute 'C. Besta', 20133 Milan, Italy; daniela.sarti@istituto-besta.it (D.S.); marinella.desalvatore@istituto-besta.it (M.D.S.); emanuela.pagliano@istituto-besta.it (E.P.); elisa.granocchio@istituto-besta.it (E.G.)

2 Department of Psychology, Catholic University of Sacred Heart, 20123 Milan, Italy; daniela.traficante@unicatt.it

* Correspondence: elisabetta.lombardi@unicatt.it

Abstract: Social distancing due to the COVID-19 pandemic represented a golden opportunity to implement telerehabilitation for clinical groups of children. The present study aims to show the impact that telerehabilitation had on the experience of well-being of children with special needs being treated at the Foundation IRCCS Neurological Institute 'C. Besta' in Milan (Specific Learning Disorders and Cerebral Palsy diagnosis); it aims to do so by comparing it with experiences of those who did not undertake telerehabilitation despite the diagnosis during the pandemic, and with typically developing children. Results show that the three groups differed in the Support, Respect and Learning dimensions of well-being experience. Post hoc comparisons revealed that children with Specific Learning Disorders and Cerebral Palsy scored higher than normotypical children in Support and in Respect scales. Furthermore, children who experienced telerehabilitation showed the highest scores on the Learning scale in comparison with the other two groups. These results support the importance of reorganizing care and assistance by integrating telemedicine, which seems to have fostered a positive experience of well-being in people with special needs, particularly in the perception of a supportive environment that respects psychological needs.

Keywords: telerehabilitation; specific learning disorders; cerebral palsy; well-being

Citation: Sarti, D.; De Salvatore, M.; Pagliano, E.; Granocchio, E.; Traficante, D.; Lombardi, E. Telerehabilitation and Wellbeing Experience in Children with Special Needs during the COVID-19 Pandemic. *Children* **2021**, *8*, 988. https://doi.org/10.3390/children8110988

Academic Editor: Sari A. Acra

Received: 21 September 2021
Accepted: 28 October 2021
Published: 1 November 2021

1. Introduction

The COVID-19 pandemic crisis is causing concern for the health and well-being of the world as a whole. Social distancing, in response to the pandemic, has also implied social isolation which has exposed, especially children, to the risk of developing relational distress and particularly, in the clinical setting, in the case of those children who were following a rehabilitation program related to their diagnosis that they have, in many cases, interrupted. The COVID-19 pandemic has intensely affected Italy, in particular Lombardy and Milan, since March 2020; among the measures implemented at national and local level to reduce the contagion, there has been the reorganization of hospitals in general. In particular, the Foundation IRCCS Neurological Institute 'C. Besta' in Milan, which had initially been conceived as a hub for non-COVID patients with neurological pathologies, in order to ensure continuity of care for all patients already followed, has reorganized and converted its activities to remote [1]. Specifically for children, this decision forced specialists to temporarily suspend all the rehabilitation protocols [2,3] necessitating alternative methods to guarantee, initially at least, the supervision of regular home exercises and activities for children with different types of special needs.

In Italy, child neuropsychiatric services were obliged to interrupt care assistance due to regulations provided by the government as measures for the containment of the virus (i.e., "Io resto a casa"—"I'm staying home" [4]). Distinct ways were followed in order to conform to these regulations: smart working, wage compensation (funds), forced holidays, part-time work. Measures such as working remotely and assured telepractice compensation from the public health system were supported by a regional decree within the Lombardy region, where Milan is the principal city, which was amongst the most affected regions. Indeed, the Italian Society of Infantile Neuropsychiatry (SINPIA) drafted a practical document about the service reorganization, highlighting the issues of important decrees for the Neuropsychiatry for Children. An important topic and worldwide line of research arose from the need to re-organize e-health clinical activities in order to ensure continuity of care for both adult and pediatric patients [5]. The Italian Institute of Health (ISS—Istituto Superiore di Sanità) has created a document [6] defining the rules and types of clinical activities within telemedicine, also including children and adolescents with special needs [7]. Moreover, international and national scientific associations have drawn up guidelines for the use of telemedicine in clinical developmental psychology and neuropsychology [8,9].

The attention provided by both clinicians and the world of research has also highlighted the risks of the pandemic emergency concerning the maintenance of human rights for children with disabilities and in poor socioeconomic conditions and disability [10]. However, there were already published studies on the different possible telerehabilitation interventions in different clinical populations available during the pre-COVID period [11]; these studies were valuable in guiding the choice of telerehabilitation courses for children with special needs, such as Cerebral Palsy, Language Disorders, Intellectual Disability, Developmental Learning Disorders. During the COVID-19 pandemic period, regarding telemedicine and in particular telerehabilitation in the pediatric age, several positive experiences are already present, some of which involve methods already in progress pre-COVID and others forcedly initiated due to the lockdown [12].

Moreover, at the Foundation IRCCS Neurological Institute 'C. Besta', the COVID-19 pandemic has provided new opportunities for rehabilitation of children with special needs, as children with Cerebral Palsy (CP) and children with Specific Learning Disorders (SLD) have benefited from the comprehensive integration of environment and family, in line with the International Classification of Functioning, Disability and Health (ICF) psychosocial model [10].

The reorganization of rehabilitation was not intended as a simple transfer of remote therapeutic paths; instead, it required the redefinition of objectives and rehabilitation methods, with the maintenance of the same therapist. In order to revise the therapeutic protocols with respect to telemedicine, the specialists verified the possibility of a child telerehabilitation basing on an informative checklist, which includes: child's characteristics (age, diagnosis, level of severity, physical and sensory aspects, cognitive-behavioral profile, communicative and linguistic skills); family compliance (access to technological device, i.e., laptop, tablet and software, adequate space at home, parents' level of collaboration, pending e-learning requests). Then, together with the parents, a new care and rehabilitation plan is developed. Several clinical centres that experimented with telemedicine and telerehabilitation pathways during the COVID-19 lockdown (first phase) have carried out research on the perception and satisfaction of stakeholders in telerehabilitation [3,13]. During the first phase of the pandemic (from March to September 2020 period) Pareyson and colleagues [1] administered a big survey to parents, caregivers and patients of the Foundation IRCCS Neurological Institute 'C. Besta' in Milan and showed that satisfaction about the telehealth and telerehabilitation was very high.

If the above underlines the impact of the pandemic on rehabilitation and care institutions, at the level of the individual, the social distancing due the pandemic undoubtedly represented a challenge also for well-being and emotional experiences for children and adolescents. This scenario led to children's perception of a feeling of isolation and lack of relational support with effect on the emotional and well-being experience, especially due to the absence of the scholastic environment [14]. From a social and ecological perspective, schools can be considered as environments in which multifaceted interactions occur at the environmental, organizational and individual level [15], together with the effect of these different levels of interactions on individual and collective health and well-being. This is not surprising, since the impact of the school context on well-being is well recognized by recent literature [16–18]. In this vein, psychological well-being is understood and studied by adopting the positive psychology perspective, namely conditions and processes that contribute to the optimal functioning of individuals, groups, and institutions [19]. Current literature begins to explore, in a multidimensional way, areas such as evolution and predictors of happiness, subjective well-being, optimism, self-determination, creativity, talent and positive youth development [20–25]; this is in line with the construct of flourishing [26–28], which extends beyond the exclusive focus on psychopathology [29]. A focus on flourishing is particularly important because childhood and adolescence are pivotal stages of development that carry lifetime implications for functioning.

The impact that COVID-19 pandemic had on the well-being of typically developing children and adolescents was immediately a research object in the most affected countries, starting with China, and then in other countries. In particular, alarming levels of behavioral and emotional disturbances in children have been detected [30,31]. The state of mental health within the pediatric population has been studied in relation to the measures adapted to contain the virus, initially analyzing the social isolation and sense of loneliness experienced [32]. This research presented revelations using investigations oriented to the psychological, emotional and social consequences of children and adolescents.

The psychological consequences of the lockdown due to the pandemic were primarily studied on caregivers, who were stressed by the fear of contagion, the activation of smart working, job losses and consequently the effects this stress had on their children. Parental self-efficacy seems to play an important role in moderating the consequences of parental distress on children's well-being [13]. Anti-contagion measures have involved not only all extracurricular activities but also schools with the start of distance learning; indeed, the consequences on children's cognitive and cultural development and well-being have already become the focus of numerous studies. Recommendations have been drawn up by scientific and political institutions to prevent educational inequalities and to propose distance learning during school closures [33–35].

What about children with special needs? Research on the impact caused by the pandemic on well-being and mental health in children with different types of special needs has been conducted through the administration of questionnaires aimed for parents, such as Child Behavior Checklist for ages 1.5–5 and for ages 4–18 [36], or questionnaires built ad hoc [37]. The most studied clinical conditions were autism, attention deficit hyperactivity disorder, obsessive compulsive disorder, disadvantaged socioeconomic status and the studies about these conditions shows the differences in emotional and behavioral profiles of children emerged on the basis of children ages, type of neurodevelopmental disorders or neurological pathology, pre-existing emotional problems, environmental factors [9,38,39]. However, to the best of our knowledge, well-being surveys have mostly involved parents, and few have involved children and adolescents.

The present study aims to show the perceived experience of well-being of children with special needs that had been through telerehabilitation during the pandemic period by comparing it with experiences of those who did not undertake telerehabilitation during the pandemic, and with typically developing children. We proposed to ask both typically developing children, and children with special needs for their direct opinion on their well-being experience. The purpose of our research is to verify the hypothesis that those who had a rehabilitation path pre-COVID and had the possibility to continue the rehabilitation process and care remotely during the emergency period, may have experienced greater well-being compared to those who had not benefited from it. Comparison with a normotypical group gives us the opportunity to consider the pandemic variable and control it. The implemented contagion containment measures, have, in fact, required an enormous adaptive effort from the entire pediatric population, greatly limiting the possibility of social relationships, the possibility of learning through direct experiences, involvement and satisfaction in the learning process, trust in future and quality of life. The emotional costs paid by children have been extremely high: caregivers, educational facilities, schools and families, who were stressed and afraid, have struggled to understand and satisfy the development needs of each child. Therefore, we expect that participants with special needs who have had a presumably better response from the environment in terms of continuity in care, rehabilitation and relationships with specialists may have had a better experience of well-being than those who have not received it.

2. Materials and Methods

2.1. Participants

In terms of participants, 56 children with different types of special needs were recruited from the clinical service of Developmental Neurology Unit of Foundation IRCCS Neurological Institute 'C. Besta' (36 children with Specific Learning Disorders and 20 children with Cerebral Palsy), and 30 normotypical children attending primary and secondary schools in Milan. All participants completed the online questionnaire, which was sent through an online survey (Google Forms) after collecting written consents by their parents, in the period from May to August 2020. All participants were native Italian speakers. The clinical study sample consisted of 36 children with Specific Learning Disorders (SLD) and 20 children with Cerebral Palsy (CP). From these two groups, children with SLD and CP who were telerehabilitation were selected for each group, respectively. These children were matched by gender, age and comorbidity in the SLD case and by age, gender and severity in the CP case with children who did not undergo telerehabilitation during the study period. More specifically, the SLD Telerehabilitation group (N = 8) had these clinical characteristics: 1 child with Dyscalculia, 2 with Dyslexia and Dysorthography, 5 with Dyslexia, Dysorthography and Dyscalculia. This group was matched with the children with SLD No telerehabilitation group (N = 8) that had the same clinical characteristics: 1 child with Dyscalculia, 2 with Dyslexia and Dysorthography, 5 with Dyslexia, Dysorthography and Dyscalculia.

For children with CP, Telerehabilitation group (N = 9) had these clinical characteristics on the base of the Classification Systems for children with Cerebral Palsy, the Gross Motor Function Classification System (GMFCS, [40]), the Manual Ability Classification System (MACS, [41]), Visual Function Classification System (VFCS, [42]): 2 children with Tetraplegia, with performances on the VFSC, GMSC and MACS between levels III and IV, with needs of substantial environmental adjustments; 4 children with Hemiplegia, with performances on VFSC, GMSC, MACS between levels I and II, with good autonomy; 3 children with Diplegia, with performances on VFSC, GMSC, MACS between levels II and III, with mild functional limitations, that need of some environmental adjustments. CP children in No telerehabilitation group (N = 9) had the same clinical characteristics: 2 subjects with Tetraplegia, 4 subjects with Hemiplegia, 3 subjects with Diplegia with clinical characteristics similar to those of the telerehabilitation group.

As for children with SLD, Telerehabilitation group (N = 8, mean age = 126.63 months; SD = 7.84 months; range min 121 months–max 145 months), had the opportunity to have 2 online treatment sessions a week (45 min), focused on reading, writing and math skills whereas children in No telerehabilitation group (N = 8, mean age = 126.62 months; SD = 7.8 months; range min 121 months– max 145 months) had finished their treatment before the pandemic period, after reaching the goals of their rehabilitation projects. More specifically they are involved in telerehabilitation with online dyslexia platform as RIDInet with Reading Trainer app and Rhythmic Reading Training RRT teleintervention [43]. The telerehabilitation used RIDInet, an internet platform that enhances reading speed and accuracy, spelling skills, text comprehension, arithmetic and numerical skills, executive functions (i.e., inhibition, working memory, cognitive flexibility), language (rapid naming, expressive skills). The Rhythmic Reading Training [43], a computer-assisted training, was designed to implement a treatment which combines a traditional approach (sublexical treatment) with rhythm processing training, and it was possible to transfer in a fairly simple and effective way remotely through the "Share screen and system audio" option present in the platforms in use the rhythmic exercises proposed.

As for children with CP, Telerehabilitation group (N = 9, mean age = 131.37 months; SD = 24.16 months; range min 93 months–max 173 months) was supported by 1 online treatment sessions (45 min), for mean 13 weeks, tailored on their neuro-psychomotor needs, whereas children in the No telerehabilitation group (N = 9, mean age = 132 months; SD = 27.8 months; range min 85 months–max 174 months) did not show, in that period, any specific need to be addressed through rehabilitation projects. Tele-treatment for children with CP consisted of real-time treatment for children with a neuropsychological and learning exercise program; sharing of either information with parents and with special need school teachers or an exercise program to be implemented by the parents or teachers. For these children with motor needs, video-tutorials were sent to parents twice a week, in which the exercises to be performed by the child were explained. The weekly supervised meeting with the therapist allowed parents to be correctly guided and above all, it allowed for subjects to maintain contact with the therapist; giving information and technical feedback about pharmacological treatment, checking the effects of the therapy on motor pattern and adaptive functions. Furthermore, two children continue the work about the study method that aimed to integrate compensatory tools in combination with the results of his CP (use of speech already started in presence with the aim of producing texts and implementing corrections independently).

All children were tested to verify the efficacy of the treatment at the end of the rehabilitation cycle. In all, a benefit was found in their performances and instrumental skills. As for the normotypical children (N = 30), recruited from school in Milan, they had no diagnosis of special needs. They were matched to the clinical groups by gender and age (see Table 1 for the descriptive statistics).

Table 1. Mean age and SD (in parentheses) of the three groups by diagnosis.

Group	N	SLD	N	CP	N	Total
Telerehabilitation	8	122.5 (13.4)	9	131.37 (24.2)	17	130.35 (30.2)
No telerehabilitation	8	126.63 (7.8)	9	132 (27.8)	17	129.47 (20.5)
Normotypical control	8	125.75 (10.7)	9	131.44 (31.4)	17	128.76 (23.5)

SLD = Specific Learning Disorder children; CP = Cerebral Palsy children.

In sum, in order to assess the impact of telerehabilitation on well-being in children with special needs, three groups were selected for both SLD and CP conditions: children with SLD or CP who made experience of telerehabilitation (N = 17; F = 8), children with SLD or CP who did not experience telerehabilitation (N = 17; F = 6), normotypical children (N = 17; F = 6). They were matched by gender ($\chi^2 = 1.416$, $p = 0.923$) and age (see Table 1; $F_{group} = 1.47$, $p = 0.232$; $F_{telerehabilitation} = 0.011$; $F_{interaction} = 0.189$).

2.2. Measures

Two questionnaires were administered to assess children's wellbeing.

2.2.1. Comprehensive Inventory of Thriving for Children

Comprehensive Inventory of Thriving for children [44] consists of a comprehensive range of subscales for assessing one facet of psychological well-being each, with three items. The Italian adaptation of the CIT to child population has 45 items assessing 15 facets of positive functioning, representing the dimensions of psychological well-being: Support (e.g., There are people who give me support and encouragement); Respect (e.g., People are polite to me); Loneliness (e.g., Often I feel left out); Belonging (e.g., I feel a sense of belonging in my community); Engagement (e.g., In most of the things I do, I feel energized); Skills (e.g., I get to do what I am good at every day); Learning (e.g., Learning new things is important to me); Self-worth (e.g., What I do in life is valuable and worthwhile); Optimism (e.g., I have a positive outlook on life); Life satisfaction (e.g., My life is going well), Positive feelings (e.g., I feel happy most of the time); Negative feelings (e.g., I feel bad most of the time). Cronbach's alpha for the total scale is 0.86.

2.2.2. Scale of Positive and Negative Experience

The Scale of Positive and Negative Experience (SPANE) [45] is a brief 12-item scale, with six items devoted to positive experiences (items: positive, good, pleasant, happy, joyful, and contented) and six items designed to assess negative experiences (items: negative, bad, unpleasant, sad, afraid, and angry). The scale is aimed to assess the full range of positive and negative experiences: it not only assesses the pleasant and unpleasant emotional feelings, but also reflects other states such as interest, flow, positive engagement, and physical pleasure. Each SPANE item is scored on a scale ranging from 1 to 5, where 1 represents "very rarely or never" and 5 represents "very often or always". Moreover, the scale is keyed to the last 4 weeks, which is short enough to allow the respondent to recall actual experiences rather than rely on general self-concept [45]. The positive and negative scales are scored separately because of the partial independence of the two types of feelings. The summed positive score (SPANE-P) can range from 6 to 30, and the negative scale (SPANE-N) has the same range. The scale has been translated into several languages, including Italian and can be downloaded for research purposes on the official website of the authors the scale has been translated into several languages, including Italian and can be downloaded for research purposes on the official website of the authors. The analysis conducted on 407 Italian children [44] showed good alpha coefficients (SPANE-P: $\alpha = 70$; SPANE-N: $\alpha = 67$).

2.3. Procedure

An invitation to participate in the research was sent to all families whose children were in charge of Developmental Neurology Unit, Foundation IRCCS Neurological Institute 'C. Besta' of Milan. The rate of positive responses was 92%. Parents who gave their consent communicated to the researchers the email address to be used to get in contact with their children. Each participant received, by email, a link to fill in the questionnaires on the online survey platform. Questionnaires were administered by researchers and were completed individually by children. Data collection lasted from May 2020 to August 2020.

2.4. Statistical Analyses

(1) In order to assess the differences in CIT-C scales between groups (Telerehabilitation, No telerehabilitation, Normotypical control), a MANOVA was carried out on the CIT-C scales raw scores with Group (Telerehabilitation, No telerehabilitation, Normotypical control) and Condition (SLD, CP) as independent variables;

(2) The effects of Group and Condition on SPANE scores (Positive and Negative) were considered as independent factors in two ANOVAs.

Post hoc comparisons were applied to analyze significant differences in detail.

3. Results

First, it is worth noting that all three groups of children (Telerehabilitation, No telerehabilitation, Normotypical control) showed mean scores on CIT-C scales within 1.5 SD from the normative mean (Table 1).

Results from MANOVA showed that the main effect of Group was significant (Group Pillai's trace: $F_{24,70} = 3.14$, $p < 001$; $\eta^2 = 0.518$), irrespective to the Condition, which did not reach significance level neither as main factor (Pillai's trace: $F_{12,34} = 1.43$, $p = 201$; $\eta^2 = 0.335$), nor in interaction with Group (Pillai's trace: $F_{24,70} = 1.51$, $p = 095$; $\eta^2 = 0.34$). One-way ANOVAs showed that the three groups differed on Support ($F_{2,45} = 7.17$, $p = 002$; $\eta^2 = 0.242$), Respect ($F_{2,45} = 3.43$, $p = 041$; $\eta^2 = 0.132$) and Learning ($F_{2,45} = 7.29$, $p = 002$; $\eta^2 = 0.245$) scales. In post hoc comparisons, Sidak adjusted (Table 2) revealed that children with SLD and CP (irrespective of the experience of Telerehabilitation) scored higher than normotypical children on the Support and in Respect scales. Children with SLD and CP, who experienced telerehabilitation, showed the highest scores on the Learning scale, in comparison with the other two groups.

Table 2. CIT-C's mean scores and SD of the three groups by diagnosis (significant differences in bold).

CIT-C Scale	Group	SLD		CP		Total	
		M	SD	M	SD	M	SD
Support	Telerehabilitation	4.46	0.5	4.89	0.2	**4.73**	0.3
	No telerehabilitation	4.71	0.3	4.74	0.4	**4.69**	0.4
	Normotypical control	4.17	1.0	3.89	0.8	**4.02**	0.9
Respect	Telerehabilitation	3.88	0.6	4.74	0.4	**4.33**	0.7
	No telerehabilitation	4.17	1.1	4.59	0.8	**4.39**	0.9
	Normotypical control	3.63	0.7	3.78	1.1	**3.71**	0.9
Loneliness	Telerehabilitation	3.50	0.5	3.52	1.0	3.51	0.8
	No telerehabilitation	4.13	0.7	3.85	0.8	3.98	0.8
	Normotypical control	4.12	0.5	3.93	1.1	4.02	0.8
Belonging	Telerehabilitation	4.29	0.9	4.44	0.9	4.37	0.9
	No telerehabilitation	4.50	0.9	4.19	1.0	4.33	0.9
	Normotypical control	4.00	0.9	3.81	0.9	3.90	0.9
Engagement	Telerehabilitation	3.21	0.8	3.89	0.9	3.57	0.9
	No telerehabilitation	3.92	0.6	3.74	0.9	3.82	0.7
	Normotypical control	3.71	0.8	2.93	0.8	3.29	0.9
Skills	Telerehabilitation	3.88	0.7	3.85	1.0	3.86	0.8
	No telerehabilitation	3.54	1.0	3.67	1.2	3.61	1.0
	Normotypical control	3.58	0.7	3.15	0.7	3.35	0.7
Learning	Telerehabilitation	4.38	0.5	4.41	0.8	**4.39**	0.7
	No telerehabilitation	3.63	0.7	3.63	0.8	**3.63**	0.8
	Normotypical control	3.79	0.8	3.30	0.4	**3.53**	0.7
Self-worth	Telerehabilitation	3.75	1.1	3.59	1.0	3.67	1.0
	No telerehabilitation	3.88	0.4	3.78	1.0	3.82	0.7
	Normotypical control	3.58	0.8	3.19	0.9	3.37	0.9
Optimism	Telerehabilitation	3.63	0.8	4.15	0.9	3.90	0.9
	No telerehabilitation	4.50	0.6	3.93	0.9	4.20	0.8
	Normotypical control	3.58	0.9	3.41	1.0	3.49	0.9
Life Satisfaction	Telerehabilitation	3.17	1.2	4.00	1.2	3.61	1.2
	No telerehabilitation	3.70	0.9	3.78	1.1	3.74	1.0
	Normotypical control	3.92	0.8	4.04	0.9	3.98	0.8
Positive feelings	Telerehabilitation	3.33	1.1	4.37	1.0	3.88	1.1
	No telerehabilitation	4.63	0.5	3.74	1.2	4.16	1.0
	Normotypical control	3.67	0.6	3.78	0.9	3.73	0.7
Negative feelings	Telerehabilitation	2.96	1.3	3.63	1.3	3.31	1.3
	No telerehabilitation	4.00	0.8	3.63	1.1	3.80	1.0
	Normotypical control	3.96	0.3	4.15	0.9	4.06	0.7

SLD = Specific Learning Disorder children; CP = Cerebral Palsy children. SPANE = Scale of Positive and Negative Experience.

ANOVAs carried out on SPANE scores did not reveal any significant difference neither by Group nor by Condition (Table 3).

Table 3. SPANE's mean scores and SD of the three groups by diagnosis.

SPANE Scale	Group	SLD		CP		Total	
		M	SD	M	SD	M	SD
Negative	Telerehabilitation	16.2	2.5	15.6	7.1	15.9	5.3
	No telerehabilitation	12.1	2.4	13.7	6.1	12.9	4.7

Table 3. *Cont.*

SPANE Scale	Group	SLD		CP		Total	
		M	SD	M	SD	M	SD
	Normotypical control	13.5	4.6	13.4	6.1	12.2	4.4
Positive	Telerehabilitation	19.1	4.0	23.9	4.7	21.7	4.9
	No telerehabilitation	25.4	2.0	22.2	6.1	23.7	4.8
	Normotypical control	22.5	5.6	23.1	5.6	22.8	5.5

SLD = Specific Learning Disorder children; CP = Cerebral Palsy children. CIT-C = Comprehensive Inventory of Thriving for children.

4. Discussion

Our research goal was to determine the effects of telerehabilitation during the pandemic period on the well-being experienced by children and adolescents with special needs. This was assessed by comparing the experiences of those who received telerehabilitation with those who did not receive it, and with typically developing children. To the best of our knowledge, this is the first study that compared the well-being of children with special needs and typical development during the lockdown phase of COVID-19 pandemic, which investigated the children's direct opinions. The questionnaires explored the psychological well-being of children from a multidimensional perspective [44,46]. Regarding these dimensions, our results reveal that children with SLD and CP, independently from the Telerehabilitation experience, scored higher on the relational dimension (support and respect scales) compared to the normotypical group. In addition, children with Specific Learning Disorders and Cerebral Palsy, who experienced telerehabilitation, showed the highest scores in Learning dimension, in comparison with the other two groups. Non-significant results are shown about the perception of negative and positive feelings.

Starting from the significant difference in Learning, the dimension between children with special needs who were in telerehabilitation during the pandemic period and children with special needs who did not is described. The Learning subscale requires to express the statement such as "new things are important to me", "I learned something new yesterday", "I always learn something every day" and it concerns a sense of mastery and accomplishment. In other words, we asked the children to answer, during the pandemic crisis, what had become of their motivation/interest in learning new subjects after being forced to stay home from school due to social distancing. In this way, the children responded about the impact and role of the environment on the pupils' intellectual curiosity. In a period of confinement due to the COVID-19 pandemic, in which remote learning has not always been reorganized in a timely manner, the perception of having learned 'new things' may have generally been limited in students. Children with special needs may have experienced, even more, the impossibility of accessing normal school learning paths and the difficult reorganization of the remote learning system, besides the development of high levels of stress/anxiety and emotional distress, in addition to low levels of well-being, self-esteem and self-efficacy [47]. The possibility for some of them to remotely continue the telerehabilitation might have influenced the well-being component linked to cultural and personal enrichment, unlike subjects with special needs who have not sustained telerehabilitation and subjects with typical development. These results help to shed a light on the role of telerehabilitation on the well-being of children with special needs. Conversely, some studies about the Specific Learning Disorders had already shown during the pre-COVID period the positive impact on the motivation of students who use new technologies [11]. Furthermore, another aspect concerns curiosity of children about the use of technological devices and its integration in the rehabilitation, which was usually carried out in person. They did not report difficulties in the use of new technologies. The one-to-one interaction and the continuous and complete attention to the child have contributed to making the intervention more dynamic and engaging despite being remotely conducted. However, it must be highlighted that remote activity requires commitment and responsibility on behalf of the child in order for it to be successful, and children therefore proved to be more active, more motivated and more responsible. The family environment setting was a strong point, also becoming a starting point for conversation. In this study, children who benefited from telerehabilitation might have felt they are improving and are generally more engaged, which may have positive impacts and effects on their well-being.

Regarding the result of the relationship dimension, our analysis shows that children with special needs, regardless of whether or not telerehabilitation has taken place, reported higher scores than children with typical development on the Respect and Support scales; these concern the perceived well-being of supportive and enriching relationships, and the environmental response component therefore appears as an important factor of well-being. The response of the Developmental Neurology Unit of Foundation IRCCS Neurological Institute 'C. Besta' of Milan since the beginning of the lockdown and the main objective of the distance reorganization, was precisely to ensure the continuity of care to patients under treatment and the provision of a prompt support response. Children with special needs in the present study were all patients already known to the specialists, with management consisting of periodic follow-ups, subsequently converted into distance sessions, and some of them were included in rehabilitation programs. For example, in line with national and international guidelines, the treatment (regular care) for children with CP is extensive, 2–3 times per week up to 6 years of age; for older aged children, depending on the objectives, the frequency of treatment is reduced (1–2 times per week), while extensive treatments are carried out in severe neuropsychological or learning disorders, and specific focus-intensive cycles, 2–3 times per year. The results of this study, even with limitations, contribute to the literature confirming that telemedicine reorganization provides a comfortable environment for young patients. Moreover, in our study, all children had already been diagnosed in our institute, although not all of them were undergoing rehabilitation. Previous research has investigated the experience of well-being in children and adolescents with special needs; indeed, the presence of neurodevelopmental disorders can be a risk factor for the development of emotional and behavioral problems, and not only has an impact on learning possibilities but also on well-being. However, some studies with students with learning difficulties show that having a diagnosis seems to have a protective role for their psychological well-being compared to those children who, despite presenting difficulties, have never received a diagnosis [17]. Recently, Lombardi and colleagues (2021) underlined the importance of receiving a diagnosis, as it seems to function as a protective agent for students' psychological and scholastic well-being. Since the two groups of children with special needs (with and without telerehabilitation) faced similar positive well-being experiences, with only one benefit in the Learning subscale in the telerehabilitation condition, it can be explained by considering the possible "costs" of telerehabilitation itself. In a difficult period for Italian families, in which parents were often engaged in smart working and distressed by feelings of fear and uncertainty, the offer of telerehabilitation—and, in general, continuity of care and assistance—may have granted children with the attention they needed from the adult world and thus filled a "void". Indeed, it is important to promote the benefits but also to assume that the commitments of telerehabilitation are often added to those of distance learning, and this may have had an impact on the well-being of those who maintained rehabilitation through telemedicine. Finally, it is important to emphasize that the three groups of subjects (Telerehabilitation, No telerehabilitation, Normotypical control) reported mean scores within 1.5 Standard Deviation from the normative mean. Furthermore, many children with special needs showed improved autonomy, and some children showed improved learning and attention during the home confinement, as confirmed by teachers. This may in part be explained by the increased time spent at home with their family or perhaps improved motivation resulting from the reduced effort and time involved in school and, for children with special needs, rehabilitation activities. It should also be noted that, with regards to the Italian adult population, [48] revealed that 38% of the general population during the COVID-19 pandemic perceived a form of psychological distress but the majority of subjects displayed no relevant distress. The timelines and quality of the reorganization in telemedicine partially filled, for some children, schools' shortcomings, which had interrupted in presence lessons and, especially in the initial phase of the pandemic, had difficulty in meeting the demands of children with special needs; this manifested as online teaching organized as lessons, whereby unpreparedness and slowness in dealing with the emergency occured.

The result regarding the positive and negative experience measured by the Scale of Positive and Negative Experience (SPANE) did not reveal any significant difference neither by Group nor by Condition. This result can be explained by the same situations and the same feelings shared by all children. The impact of the restrictions due to the pandemic involved the whole world and led to shared feelings, regardless of condition. This has led to children answering questions such as "in the last 4 weeks I have felt sad" or "in the last 4 weeks I have felt happy" in the same way, regardless of whether they are in telerehabilitation or a typically developing child.

In conclusion, our findings support the importance of having reorganized care and assistance by integrating telemedicine, which may have fostered a positive experience of well-being in people with special needs, particularly in the perception of a supportive environment that respects psychological needs. The aim was to ensure continuity of care, while remaining aware that this could not replace in-person interventions. Home-based rehabilitation is an emerging feature in the literature and the difficulty, which has not always been overcome, has been that of transferring the complex regimes based on motor learning principles and psychological consultation to the home setting. The possibility of maintaining remote rehabilitation also seems to have nurtured feelings of mastery and accomplishment in children with special needs. We would therefore like to see the maintenance of hybrid methods, integrating telemedicine into clinical practice, adapted to the needs of each individual and with attention to the psychological well-being of young patients.

5. Limitations

Although the results of our study give some insights about the opportunity for telerehabilitation to support students' well-being, some limitations should be emphasized, concerning the size of the sample and the not entirely homogeneous nature of the types of treatments and clinical profiles. A larger sample size would have allowed us to consider the different types of treatment performed and also to analyze the relationships and comparisons between the different variables through more complex statistical techniques, which would allow for a greater generalization of results. This was also due to the period in which the data were collected, and the initial phase of the reorganization, which prompted an investigation into these issues, regardless. However, to the best of our knowledge, this is the first study to ask children with special needs how they were approaching this period of difficulty using a control group with typical development. Therefore, we hope that the present study might be considered a "small part" of the literature on the topic of telemedicine with a focus on its effects on the well-being of young patients.

Author Contributions: Conceptualization, D.S., E.L., D.T., M.D.S.; methodology, E.L., D.T.; formal analysis, E.L., D.T. Investigation, M.D.S., E.P.; data curation, M.D.S., D.S., E.L., E.P.; writing—original draft preparation, D.S., M.D.S.; writing—review and editing, E.L., D.T., E.P., E.G. All authors have read and agreed to the published version of the manuscript.

Funding: This research received no external funding.

Institutional Review Board Statement: The study was conducted according to the guidelines of the Declaration of Helsinki, and approved by the Ethics Committee of the Foundation IRCCS Neurological Institute 'C. Besta', Milan, Italy (protocol code: fonDL Vers. 2.0 del 20 June 2020).

Informed Consent Statement: Parents provided written consent for their children's participation in the study and students gave informed written consent to the study, according to the General Data Protection Regulation (GDPR 2016/79, 25 May 2018).

Data Availability Statement: The data presented in this study are available on request from the authors. The data are not publicly available due to privacy restrictions.

Acknowledgments: We are grateful to the families and children for participating in the study. We thank Sara Magenes for help in recruiting the normotypic sample.

Conflicts of Interest: The authors declare no conflict of interest.

References

1. Pareyson, D.; Pantaleoni, C.; Eleopra, R.; De Filippis, G.; Moroni, I.; Freri, E.; Zibordi, F.; Bulgheroni, S.; Pagliano, E.; Sarti, D.; et al. Neuro-telehealth for fragile patients in a tertiary referral neurological institute during the COVID-19 pandemic in Milan, Lombardy. *Neurol. Sci.* **2021**, *42*, 2637–2644. [CrossRef]
2. Sarti, D.; De Salvatore, M.; Gazzola, S.; Pantaleoni, C.; Granocchio, E. So far so close: An insight into smart working and telehealth reorganization of a Language and Learning Disorders Service in Milan during COVID-19 pandemic. *Neurol. Sci.* **2020**, *41*, 1659–1662. [CrossRef] [PubMed]
3. Taddei, M.; Bulgheroni, S. Facing the real time challenges of the COVID-19 emergency for child neuropsychology service in Milan. *Res. Dev. Disabil.* **2020**, *107*, 103786. [CrossRef] [PubMed]
4. Italian Government: Measures to Face the Coronavirus COVID-19. 2020. Available online: http://www.governo.it/it/coronavirus (accessed on 20 August 2021).
5. Maresca, G.; Maggio, M.G.; De Luca, R.; Manuli, A.; Tonin, P.; Pignolo, L.; Calabrò, R.S. Tele-Neuro-Rehabilitation in Italy: State of the Art and Future Perspectives. *Front. Neurol.* **2020**, *11*, 11. [CrossRef] [PubMed]
6. Italian Institute of Health. Rapporto ISS COVID-19 n. 12/2020—Indicazioni ad interim per servizi assistenziali di telemedicina durante l'emergenza sanitaria COVID-19. Versione del 13 aprile 2020. 2020. Available online: https://www.iss.it/rapporti-covid-19/-/asset_publisher/btw1J82wtYzH/content/Rapporto-iss-covid-19-n.-12-2020-indicazioni-ad-interim-per-servizi-assistenziali-di-telemedicina-durante-l-emergenza-sanitaria-covid-19 (accessed on 21 August 2021).
7. Stipa, G.; Gabbrielli, F.; Rabbito, C.; Di Lazzaro, V.; Amantini, A.; Grippo, A.; Carrai, R.; Pasqui, R.; Barloscio, D.; Olivi, D.; et al. The Italian technical/administrative recommendations for telemedicine in clinical neurophysiology. *Neurol. Sci.* **2021**, *42*, 1923–1931. [CrossRef]
8. Fazzi, E.; Castelli, E. Presentation of recommendations for the rehabilitation of children with cerebral palsy. *Eur. J. Phys. Rehabil. Med.* **2015**, *52*, 691–703.
9. Conti, E.; Sgandurra, G.; De Nicola, G.; Biagioni, T.; Boldrini, S.; Bonaventura, E.; Buchignani, B.; Della Vecchia, S.; Falcone, F.; Fedi, C.; et al. Behavioural and Emotional Changes during COVID-19 Lockdown in an Italian Paediatric Population with Neurologic and Psychiatric Disorders. *Brain Sci.* **2020**, *10*, 918. [CrossRef]
10. Schiariti, V.; Mâsse, L. Identifying relevant areas of functioning in children and youth with Cerebral Palsy using the ICF-CY coding system: From whose perspective? *Eur. J. Paediatr. Neurol.* **2014**, *18*, 609–617. [CrossRef] [PubMed]
11. Camden, C.; Pratte, G.; Fallon, F.; Couture, M.; Berbari, J.; Tousignant, M. Diversity of practices in telerehabilitation for children with disabilities and effective intervention characteristics: Results from a systematic review. *Disabil. Rehabil.* **2019**, *42*, 1–13. [CrossRef]
12. Fazzi, E.; Galli, J. New clinical needs and strategies for care in children with neurodisability during COVID-19. *Dev. Med. Child Neurol.* **2020**, *62*, 879–880. [CrossRef]
13. Assenza, C.; Catania, H.; Antenore, C.; Gobbetti, T.; Gentili, P.; Paolucci, S.; Morelli, D. Continuity of Care During COVID-19 Lockdown: A Survey on Stakeholders' Experience With Telerehabilitation. *Front. Neurol.* **2021**, *11*, 617276. [CrossRef]
14. Ammaniti, M. *E poi i bambini: I nostri figli al tempo del Coronavirus*; Solferino: Solferino, Italy, 2020.
15. Dooris, M. Holistic and sustainable health improvement: The contribution of the settings-based approach to health promotion. *Perspect. Public Health* **2009**, *129*, 29–36. [CrossRef]
16. Lombardi, E.; Traficante, D.; Bettoni, R.; Offredi, I.; Vernice, M.; Sarti, D. Comparison on Well-Being, Engagement and Perceived School Climate in Secondary School Students with Learning Difficulties and Specific Learning Disorders: An Exploratory Study. *Behav. Sci.* **2021**, *11*, 103. [CrossRef]
17. Lombardi, E.; Traficante, D.; Bettoni, R.; Offredi, I.; Giorgetti, M.; Vernice, M. The Impact of School Climate on Well-Being Experience and School Engagement: A Study with High-School Students. *Front. Psychol.* **2019**, *10*, 2482. [CrossRef] [PubMed]
18. Wang, M.-T.; DeGol, J.L. School Climate: A Review of the Construct, Measurement, and Impact on Student Outcomes. *Educ. Psychol. Rev.* **2016**, *28*, 315–352. [CrossRef]
19. Gable, S.L.; Haidt, J. What (and Why) is Positive Psychology? *Rev. Gen. Psychol.* **2005**, *9*, 103–110. [CrossRef]
20. Buss, D.M. The evolution of happiness. *Am. Psychol.* **2000**, *55*, 15–23. [CrossRef]
21. Larson, R.W. Toward a psychology of positive youth development. *Am. Psychol.* **2000**, *55*, 170. [CrossRef] [PubMed]
22. Diener, E. Subjective well-being: The science of happiness and a proposal for a national index. *Am. Psychol.* **2000**, *55*, 34–43. [CrossRef] [PubMed]
23. Peterson, C. The future of optimism. *Am. Psychol.* **2000**, *55*, 44–55. [CrossRef] [PubMed]
24. Ryan, R.M.; Deci, E. Intrinsic and Extrinsic Motivations: Classic Definitions and New Directions. *Contemp. Educ. Psychol.* **2000**, *25*, 54–67. [CrossRef] [PubMed]

25. Simonton, D.K. Creativity: Cognitive, personal, developmental, and social aspects. *Am. Psychol.* **2000**, *55*, 151–158. [CrossRef]
26. Keyes, C.L.M. The Mental Health Continuum: From Languishing to Flourishing in Life. *J. Health Soc. Behav.* **2002**, *43*, 207–222. [CrossRef]
27. Keyes, C.L.M. Mental Illness and/or Mental Health? Investigating Axioms of the Complete State Model of Health. *J. Consult. Clin. Psychol.* **2005**, *73*, 539–548. [CrossRef] [PubMed]
28. Seligman, M.E. Flourish: A visionary new understanding of happiness and well-being. *Policy* **2011**, *27*, 60–61.
29. Seligman, M.E.P.; Csikszentmihalyi, M. Positive psychology: An introduction. *Am. Psychol.* **2000**, *55*, 5–14. [CrossRef] [PubMed]
30. Liu, J.J.; Bao, Y.; Huang, X.; Shi, J.; Lu, L. Mental health considerations for children quarantined because of COVID-19. *Lancet Child Adolesc. Health* **2020**, *4*, 347–349. [CrossRef]
31. Jiao, W.Y.; Wang, L.N.; Liu, J.; Fang, S.F.; Jiao, F.Y.; Pettoello-Mantovani, M.; Somekh, E. Behavioral and Emotional Disorders in Children during the COVID-19 Epidemic. *J. Pediatr.* **2020**, *221*, 264–266.e1. [CrossRef]
32. Loades, M.E.; Chatburn, E.; Higson-Sweeney, N.; Reynolds, S.; Shafran, R.; Brigden, A.; Linney, C.; McManus, M.N.; Borwick, C.; Crawley, E. Rapid Systematic Review: The Impact of Social Isolation and Loneliness on the Mental Health of Children and Adolescents in the Context of COVID-19. *J. Am. Acad. Child Adolesc. Psychiatry* **2020**, *59*, 1218–1239.e3. [CrossRef]
33. Goldschmidt, K. The COVID-19 Pandemic: Technology use to Support the Wellbeing of Children. *J. Pediatr. Nurs.* **2020**, *53*, 88–90. [CrossRef]
34. Zhang, W.; Wang, Y.; Yang, L.; Wang, C. Suspending Classes Without Stopping Learning: China's Education Emergency Management Policy in the COVID-19 Outbreak. *J. Risk Financ. Manag.* **2020**, *13*, 55. [CrossRef]
35. Bayrakdar, S.; Guveli, A. *Inequalities in Home Learning and Schools' Provision of Distance Teaching during School Closure of COVID-19 Lockdown in the UK*; ISER Working Paper Series; University of Essex: Colchester, UK, 2020.
36. Achenbach, T.M. Child Behavior Checklist. In *Encyclopedia of Clinical Neuropsychology*; Kreutzer, J.S., DeLuca, J., Caplan, B., Eds.; Springer: New York, NY, USA, 2011. [CrossRef]
37. Cantiani, C.; Dondena, C.; Capelli, E.; Riboldi, E.; Molteni, M.; Riva, V. Effects of COVID-19 Lockdown on the Emotional and Behavioral Profiles of Preschool Italian Children with and without Familial Risk for Neurodevelopmental Disorders. *Brain Sci.* **2021**, *11*, 477. [CrossRef] [PubMed]
38. Masi, A.; Diaz, A.M.; Tully, L.; Azim, S.I.; Woolfenden, S.; Efron, D.; Eapen, V. Impact of the COVID-19 pandemic on the well-being of children with neurodevelopmental disabilities and their parents. *J. Paediatr. Child Health* **2021**, *57*, 631–636. [CrossRef]
39. Bentenuto, A.; Mazzoni, N.; Giannotti, M.; Venuti, P.; de Falco, S. Psychological impact of Covid-19 pandemic in Italian families of children with neurodevelopmental disorders. *Res. Dev. Disabil.* **2021**, *109*, 103840. [CrossRef] [PubMed]
40. Palisano, R.; Rosenbaum, P.; Walter, S.; Russell, D.; Wood, E.; Galuppi, B. Development and reliability of a system to classify gross motor function in children with cerebral palsy. *Dev. Med. Child Neurol.* **2008**, *39*, 214–223. [CrossRef]
41. Eliasson, A.-C.; Krumlinde-Sundholm, L.; Rösblad, B.; Beckung, E.; Arner, M.; Öhrvall, A.-M.; Rosenbaum, P. The Manual Ability Classification System (MACS) for children with cerebral palsy: Scale development and evidence of validity and reliability. *Dev. Med. Child Neurol.* **2006**, *48*, 549–554. [CrossRef] [PubMed]
42. Baranello, G.; Signorini, S.; Tinelli, F.; Guzzetta, A.; Pagliano, E.; Rossi, A.; Foscan, M.; Tramacere, I.; Romeo, D.M.M.; Ricci, D.; et al. Visual Function Classification System for children with cerebral palsy: Development and validation. *Dev. Med. Child Neurol.* **2020**, *62*, 104–110. [CrossRef]
43. Bonacina, S.; Cancer, A.; Lanzi, P.L.; Lorusso, M.L.; Antonietti, A. Improving reading skills in students with dyslexia: The efficacy of a sublexical training with rhythmic background. *Front. Psychol.* **2015**, *6*, 1510. [CrossRef]
44. Andolfi, V.R.; Tay, L.; Traficante, D. Assessing well-being in children: Italian adaptation of the Comprehensive Inventory of Thriving for Children (CIT-Child). *Test. Psychom. Methodol. Appl. Psychol.* **2017**, *24*, 127–145.
45. Diener, E.; Wirtz, D.; Biswas-Diener, R.; Tov, W.; Kim-Prieto, C.; Choi, D.-W.; Oishi, S. New Measures of Well-Being. In *Social Assessing Well-Being*; Springer: Dordrecht, The Netherlands, 2009; pp. 247–266.
46. Su, R.; Tay, L.; Diener, E. The Development and Validation of the Comprehensive Inventory of Thriving (CIT) and the Brief Inventory of Thriving (BIT). *Appl. Psychol. Health Well-Being* **2014**, *6*, 251–279. [CrossRef]
47. Cataudella, S.; Carta, S.; Mascia, M.; Masala, C.; Petretto, D.; Agus, M.; Penna, M. Teaching in Times of the COVID-19 Pandemic: A Pilot Study on Teachers' Self-Esteem and Self-Efficacy in an Italian Sample. *Int. J. Environ. Res. Public Health* **2021**, *18*, 8211. [CrossRef] [PubMed]
48. Moccia, L.; Janiri, D.; Pepe, M.; Dattoli, L.; Molinaro, M.; De Martin, V.; Chieffo, D.; Janiri, L.; Fiorillo, A.; Sani, G.; et al. Affective temperament, attachment style, and the psychological impact of the COVID-19 outbreak: An early report on the Italian general population. *Brain Behav. Immun.* **2020**, *87*, 75–79. [CrossRef] [PubMed]

Article

Towards Consensus on Good Practices for the Use of New Technologies for Intervention and Support in Developmental Dyslexia: A Delphi Study Conducted among Italian Specialized Professionals

Maria Luisa Lorusso [1,*], Francesca Borasio [1,2], Martina Da Rold [3] and Andrea Martinuzzi [3]

[1] Unit of Child Psychopathology, Scientific Institute IRCCS E. Medea, 23842 Bosisio Parini, Italy; francesca.borasio@lanostrafamiglia.it
[2] Department of Psychology, Catholic University of the Sacred Heart, 20123 Milano, Italy
[3] Epilepsy and Neurophysiology Unit, Scientific Institute IRCCS E. Medea, 31015 Conegliano, Italy; martina.darold@lanostrafamiglia.it (M.D.R.); andrea.martinuzzi@lanostrafamiglia.it (A.M.)
* Correspondence: marialuisa.lorusso@lanostrafamiglia.it

Abstract: The use of new technologies for intervention in developmental dyslexia is steadily growing. In order to better understand the needs, the expectations, and the attitudes of Italian expert health professionals concerning such technologies, a national survey was conducted applying the Delphi methodology. Ad-hoc questionnaires were sent out to a group of eighteen experts over three successive rounds, and anonymously collected responses were aggregated and shared with the group after each round, aiming to reach a consensus on the proposed response. The goal was to define a series of statements that could form the basis for international "good practices" in the use of technologies for intervention to support dyslexia in children and adolescents. In the first round, the experts' general opinions were collected with both multiple choice and open questions, and in the second round consensus was assessed on a series of statements based on the first replies. The cut-off of 75% consensus on each statement was reached after three rounds. Fifteen experts completed all the rounds of the process, and a final version of the statements regarding good practice in the use of technologies for dyslexia could be defined.

Keywords: developmental dyslexia; new technologies; intervention; augmented reality; virtual reality; good practices; Delphi method; rehabilitation

Citation: Lorusso, M.L.; Borasio, F.; Da Rold, M.; Martinuzzi, A. Towards Consensus on Good Practices for the Use of New Technologies for Intervention and Support in Developmental Dyslexia: A Delphi Study Conducted among Italian Specialized Professionals. *Children* **2021**, *8*, 1126. https://doi.org/10.3390/children8121126

Academic Editor: Jareen K. Meinzen-Derr

Received: 30 September 2021
Accepted: 22 November 2021
Published: 3 December 2021

1. Introduction

Recent decades have seen a change in educational and clinical settings under the influence of information and communications technology (ICT). ICTs are technological systems (e.g., hardware devices and software applications) that allow the production, storage, communication, and sharing of information [1,2]. The various types of technology currently used in education and clinical activity meet the diverse needs of the users, from teachers and clinicians or parents, to students and patients. Categorization into low-tech, mid-tech, and high-tech is based on the usability, development, and practicality of the technologies; it ranges from highlighters or adapted pencils, to computers, mobile devices, and artificial intelligence applications [3,4].

In recent years, high-tech technologies were adopted in studies regarding learning disorders, for instance text-to-speech and speech-to-text applications for the empowerment of reading and writing skills [5]. Technologies may support the learning process, giving more opportunity to practice, providing immediate feedback and an individualized and flexible learning environment [4,6].

Augmented reality and virtual reality are two types of technology used in education and intervention projects.

Augmented reality (AR) provides an interactive experience with the real world, where objects in the world are enhanced by perceptual information generated by the computer [7]. The first applications of AR in the education setting, designed twenty-five years ago, were mainly based on head-mounted displays and were too expensive and complex to be used in extensive field experiments. The second generation of AR, from 2010 to 2019, was mainly integrated in mobile devices, with the start of a rather controversial research line focused on the use of smart glasses, attempting to empower people's lives by placing information just before their eyes. These changes resulted in reduced costs and increased usability. The third generation of the present day is further enriched by artificial intelligence (AI) and AR-based web, providing various types of users with more autonomy, social integration, motivation, and enjoyment [4]. Studies have shown that AR may offer several advantages when used in educational settings [7,8]. For instance, AR helps university students build laboratory skills and positive attitudes relating to physics laboratories [9], in addition to aiding reading comprehension tasks through motivating and interesting games that promote problem solving, exploration, and socialization [10].

Virtual reality (VR) is an interactive simulation created with computer hardware and software to generate a fully immersive experience, an environment that appears similar to the real world [11,12]. There are different types of VR that range from fully immersive to non-immersive according to the degree of isolation from the physical surroundings while the user interacts with the virtual environment [13]. In recent years, VR has been used in the context of health care and has been proven to represent an alternative possibility for neurorehabilitation [14–16]. Similarly to AR, VR also allows the creation of tailor-made training programs and the adaptation of the rehabilitation process to each patient's specific needs [14].

Several studies underlined the effects of technology on students with and without special needs, such as students with learning disorders [5,6,17].

Dyslexia is a neurodevelopmental disorder characterized by difficulties in accurate or fluent word recognition and spelling, which occurs despite typical intelligence, adequate educational resources, motivation, and an absence of neurological or psychiatric problems [18,19].

Developmental dyslexia, according to national clinical guidelines [20,21], is diagnosed from the end of the second year of schooling (when children are between seven and eight years old) and appropriate treatment can be provided at either public or private centers. Nonetheless, intervention may be provided even before that age in the case of known risk factors (e.g., family history or concomitant language disorders with phonological impairment) or in the prospect of prevention. The guidelines also specify that treatment can be intensive and should last between a few weeks to three/four months, possibly provided in subsequent cycles [20,21].

Some studies have pointed out that the treatment of dyslexia may require intensive training, explicit instructions for the exercises, and single person or small group implementation [22,23]. Standard treatment methods use a classic paper and pencil format, but children may find such exercises boring or too repetitive. Technology represents great support for dyslexic children in order to achieve their educational goals [6,16,17,24]. A recent review has reported improvements in reading, writing, and calculation skills following the use of technological tools in children with learning disorders [5].

Various studies using Wii-based, or computer, games have shown that a short and intensive treatment with an action-based video game, rather than the training of phonological or orthographic skills, may improve reading abilities in dyslexic children, specifically visual-attention abilities, spatial cognition, auditory spatial attention, and response time speed [25–29]. A recent study using VR technology for the rehabilitation of reading deficits in dyslexia found that a virtual environment may represent a valid approach to improve attentional skills [16].

The applicability, usability, and practicality of these technological tools are important elements to check when deciding to implement a technological treatment program. Moreover, in order to produce positive effects in reading and writing skills in children with dyslexia, it is crucial to tailor technological tools to the specific characteristics of children. When designing ICT tools, one should take into consideration, for example, font types and sizes or screen colors that facilitate reading [5,30]. Even though all of the ICT tools are considered to be useful, there are still few studies on the use and effects of artificial intelligence, augmented reality, and virtual reality [6].

The main goal of the present study is to formulate a series of statements that could form the basis for international "good practices" on the use of technologies in the treatment of dyslexia or specific reading disorders in children and adolescents. The study is situated in the context of the European ForDys-Var Erasmus+ project (https://fordysvar.eu/ (accessed on 20 November 2021)), whose objective is to improve learning in people with dyslexia through technology, specifically virtual reality and augmented reality. The project is being conducted in collaboration with three different European countries, namely Italy, Romania, and Spain.

During the present study, an online survey was conducted in order to reach a consensus on recommendations using the Delphi method. The Delphi method, originally created in the 1950s by the Rand corporation [31], is a group facilitation process aiming to obtain a consensus regarding the opinions of experts through multiple rounds of questionnaires. After each round, anonymous responses are aggregated and shared with the group. Before starting such a multistage process in order to combine opinion into group consensus, a panel of carefully selected participants must be identified. This group of experts should demonstrate involvement and expertise in the field related to the research topic [31,32]. Although it is crucial to have adequate panel members to form a heterogeneous group, there is no agreement regarding the optimal size of a Delphi panel, with several studies including fewer than 20 participants (e.g., [33,34]). The panel of experts receives an initial Delphi questionnaire that may include open-ended questions, qualitative comments are encouraged. Then, comments from the whole group are sent to the participants through a second questionnaire, and feedback is requested to show the comparison between the individual's ratings and the whole distribution. Statements can be modified considering the feedback and a third questionnaire is thus formulated. This process is repeated until an adequate degree of consensus is reached among the group.

2. Materials and Methods

2.1. Questionnaire Design

Following Murphy and colleagues [35], a three-round Delphi survey was conducted for this study. In particular, we used the digital method, called the e-Delphi method, consisting of an online survey platform to collect data [36]. An agreement of at least 75% on each question was proposed to define a consensus. The questionnaire included both statements on the use of ICT in general, and more specific questions on AR and VR, as the study aimed to collect opinion and consensus on the general use of technology for intervention addressing reading disorders. Specifically, VR and AR were chosen to represent some of the most cutting-edge technologies, which were also developed and employed in the ForDysVar project.

2.2. Participants

The online questionnaire was sent to a group of 18 professionals, including 13 psychologists, 3 child neuropsychiatrists, and two speech-and-language pathologists who are among the most recognized experts in Italy in the field of intervention for dyslexia and who were known by the authors to have at least some experience with intervention tools based on new technologies. All of the selected experts are part of at least one of the main Italian scientific associations involved in the study and clinical practice of reading disorders: AIRIPA (Associazione Italiana per la Ricerca e l'Intervento in Psicopatologia dell'Apprendimento—Italian Society for Research and Intervention in the Psychopathology of Learning processes) and AID (Associazione Italiana Dislessia—Italian Dyslexia

Association). Some of the experts had personally contributed as scientific consultants to the development and/or validation (not-for-profit) of technological tools for intervention in neurodevelopmental disorders (mainly computerized games aimed to improve phonological skills, memory, or attention). None of them had any conflict of interest. The experts were invited by email. In the email text, the title, and the instructions of the online questionnaire, it was clearly stated that the goal of the survey was to define a set of international "good practices" for the use of technologies for the treatment and support of developmental dyslexia.

2.3. Procedure

2.3.1. Data Collection

Data from the three rounds of the e-Delphi survey were collected between September 2020 and February 2021. Before starting the online survey, participants were informed (both in an email and in the online questionnaire) that their responses would be recorded in a completely anonymous form with no possibility to retrieve the respondents' identities. They were further informed that the completion of the questionnaire implied that they agreed to the collection and the processing of their responses in this anonymous form, as well as to their use for scientific purposes and future publications. All the questionnaires were implemented using the Google Form application, with obligatory responses to each of the closed questions. For each round, three emails were sent to the whole panel over a period of 2–3 weeks, and the collection of the responses was closed 3 weeks after the last email. During each round, the experts received the link to access the results from the previous round as reported on the Google Form page (after response collection had been closed).

2.3.2. Round 1

The questionnaire for Round 1 consisted of 21 questions about technology applied to dyslexia (12 multiple choice questions and nine open-ended questions, see Table 1). The questions and the response options were formulated based on previous literature as to represent the most controversial issues for clinical use. Since the literature did not always provide specific information, some of the questions were based on the authors' direct clinical experience with technology for the rehabilitation of reading disorders, or on their own opinions, always providing answers that could either confirm or deny their hypotheses. The panel could provide comments and suggestions for each of the questions (an open-ended question for comments was provided after each of the question in the online survey form). At the end of this stage, the replies were analyzed in order to formulate the statements that were to be rated by the same group of experts in the second step of the Delphi procedure.

Table 1. Round 1 questions and response options.

Questions	Answers
(1) In your opinion, can ICT technology support the treatment of Dyslexia?	-Yes, I believe it could play a preeminent role compared to other methods of treatment -Yes, I believe it could be as good as other methods of treatment -Yes, but not as significant as other methods -No
(2) Do you know any systems based on ICT technologies applied to the rehabilitation of SLDs, specific learning disorders, in particular, dyslexia?	-Yes, I currently use them in clinical practice -Yes, but I do not use them -No
(3) What kind of software/systems did you use?	Open-ended question

Table 1. *Cont.*

Questions	Answers
(4) In your opinion, what are the advantages of using ICT tools for the treatment of dyslexia? (You can choose more than one answer)	-Easy to use -The opportunity to be completed daily and several times per week -Cost-effectiveness -The practicality of being carried out at different times of the day or in different environments (at home, at school etc.) -It is more motivating/engaging
(5) Do you consider the treatment of dyslexia more effective with software that enhances? (you can choose more than one answer)	-Grapheme–phoneme conversion processes -Assembly processes of the phonological structure -Lexical processes -Visual analysis processes
(6) In your opinion, what is the ideal duration of treatment carried out with ICT tools?	-A month -2 to 3 months -3 to 6 months -More than 6 months
(7) At what age do you think it is more appropriate to start treatment using ICT tools?	-Before the start of primary school -First two years of primary school -From the third year of primary school -Middle school -High school
(8) In your opinion, does the use of ICT in rehabilitation support the motivation to learn?	-Yes -No -I am skeptical
(9) In your opinion, can augmented reality be used to create treatment tools for children and/or teenagers with dyslexia?	-Yes -No -I am skeptical
(10) If yes, how?	Open-ended question
(11) If yes, from what age?	Open-ended question
(12) If yes, with what aim?	Open-ended question
(13) In your opinion, can virtual reality be used to create treatment tools for children and/or teenagers with dyslexia?	-Yes -No -I am skeptical
(14) If yes, how?	Open-ended question
(15) If yes, from what age?	Open-ended question
(16) If yes, with what aim?	Open-ended question
(17) What limits do you see in the use of ICT tools for the treatment of dyslexia?	Open-ended question
(18) In your opinion, can ICT tools facilitate the learning of school content in children and/or teenagers with dyslexia?	-Yes -No -I am skeptical

Table 1. *Cont.*

Questions	Answers
(19) If yes, how do you imagine the proposal for an ICT-based learning activity?	Open-ended question
(20) Do you think that virtual reality is suitable for this purpose?	-Yes -No -I am skeptical -I don't know
(21) Do you think that augmented reality is suitable for this purpose?	-Yes -No -I am skeptical -I don't know

2.3.3. Round 2

After the completion of Round 1, 39 statements based on the previous survey were sent to the same group of experts. Open space was added after each statement to suggest possible improvements. The experts were asked to express their degree of agreement with each statement on a 5-level, Likert-like scale from 1 (strongly disagree) to 5 (strongly agree). The intermediate level 3 (labelled as "I do not know") was used to express a lack of knowledge or expertise. An overall 75% group consensus was the target required to determine a positive outcome and stop the process.

2.3.4. Round 3

The questionnaire was further revised following Round 2, providing alternative wordings for the statements that had not reached the consensus cut-off of 75% in Round 2. Participants were requested to express their degree of agreement on the same scale used in Round 1, with the new statements only. This was sent to all panel members and their replies were collected. Figure 1 shows the flowchart of the Delphi process.

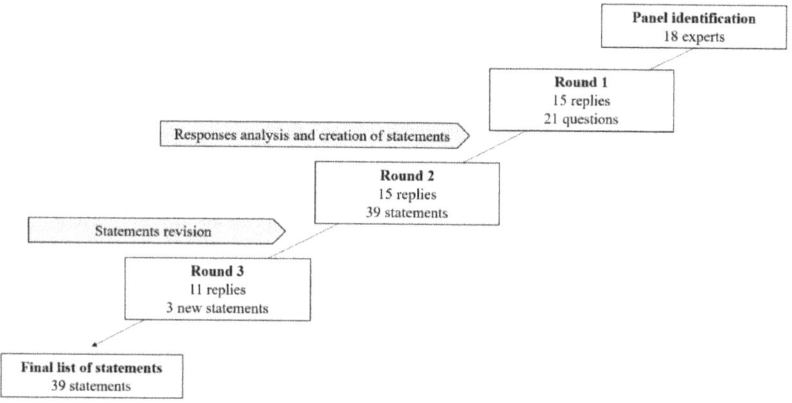

Figure 1. Flowchart of the Delphi process.

2.4. Data Analysis

Data collected in the three rounds of the Delphi survey were analyzed qualitatively. Questions of Round 2 and Round 3 provided answers that could be led back to an ordinal scale with five levels, possible replies were "strongly disagree, disagree, do not know, agree, and strongly agree". The answer "I do not know" was initially assigned an intermediate value of 3 in equivalent scores, but when it was observed (from the open-ended responses) that its use was limited to experts who declared to have no or little knowledge of the device being judged, it was decided to consider it as a null reply and not to include it in the calculation of the degree of agreement. Thus, the scale was treated as a four-level scale, with the possible answers "strongly disagree, disagree, agree, and strongly agree". Agreement was thus calculated as the percentage of the ratings above 3 (4 = agree, 5 = strongly agree) on the total number of responses, excluding 3 (= I do not know).

3. Results

3.1. Round 1

The responses collected in the first survey are presented below. Fifteen experts out of 18 completed the survey. All the respondents declared that information and communications technology (ICT) can support the treatment of dyslexia. In particular, 46.7% indicated that it could be as good as other methods, 40% indicated that it could play a preeminent role compared to other intervention methods, and 13% declared that its contribution could not be as significant as other methods (Figure 2a).

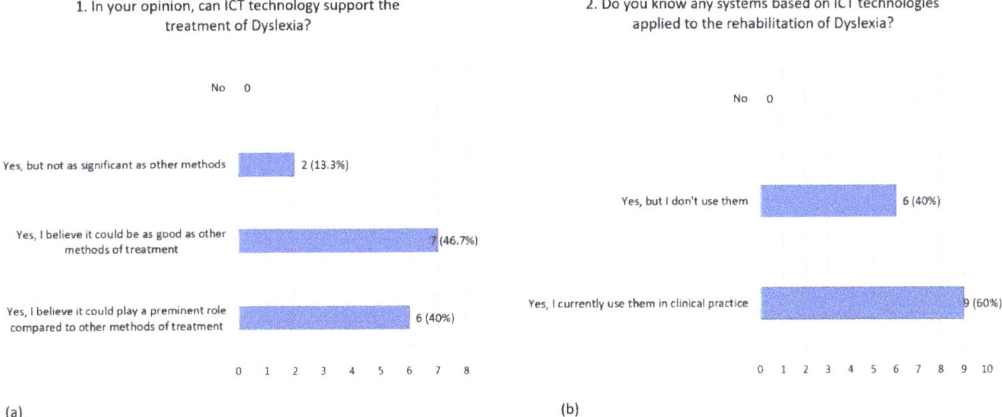

Figure 2. Distribution of responses to question 1 (**a**) and question 2 (**b**) on information and communications technology (ICT).

The experts declared to know some systems based on ICT technologies applied on the rehabilitation of dyslexia A total of 60% would use them in clinical practice, and 40% would not use them (Figure 2b). Specifically, experts using ICT are familiar with different kinds of software and systems widely used in Italy, and other software involving the stimulation of phonological, lexical, sublexical reading-related processes, visual recognition of graphemes, spelling, and memory.

The advantage of using ICT tools for intervention in dyslexia is considered to lie in the ease of use (46.7%), the possibility of an intensive use (100%), cost-effectiveness (46.7%), the possibility to use them in different environments and at different times of the day (73.3%), and their motivating and engaging characteristics (46.7%) (Figure 3a). The treatment of dyslexia is considered to be more effective if based on software improving phonological assembly processes (66.7%), lexical processes (53.3%), visual analysis processes (53.3%), and grapheme–phoneme conversion processes (46.7%) (Figure 3b).

As to the question concerning optimal treatment duration, 46.7 % believe that the ideal treatment duration is from 2 to 3 months, 46.7% from 3 to 6 months, and only 6.6% indicated a one-month-duration (Figure 4a). The most appropriate age to start a treatment using ICT tools was considered to be during the first two years of primary school (66.7%) or from the third year of primary school on (33.3%) (Figure 4b).

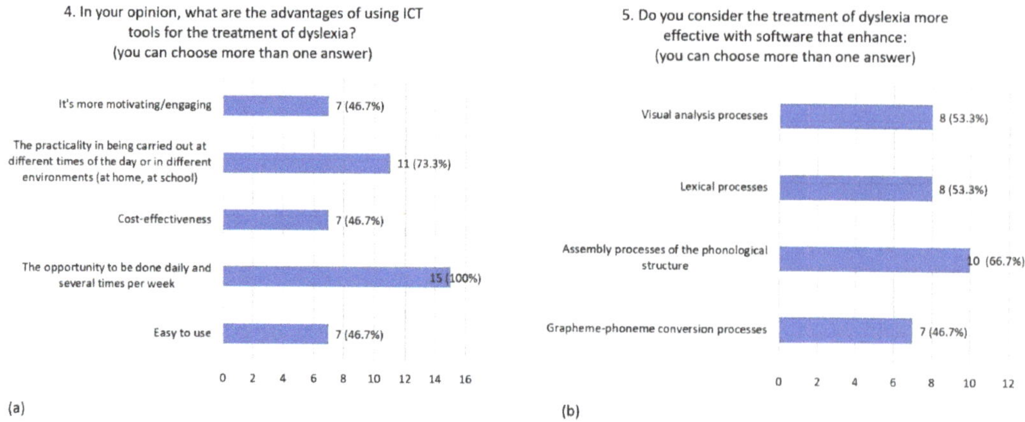

Figure 3. Distribution of responses to question 4 (**a**) and question 5 (**b**).

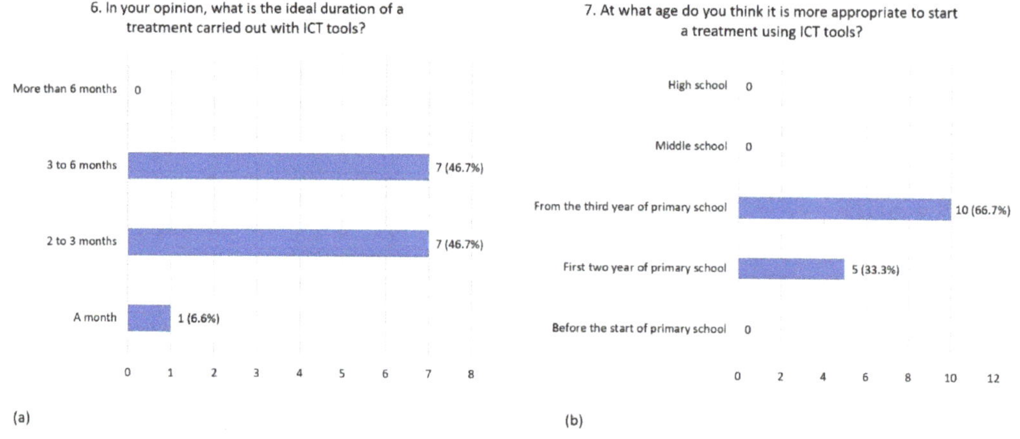

Figure 4. Distribution of responses to question 6 (**a**) and question 7 (**b**).

Almost all respondents declared that the use of ICT tools in treatment supports the child's motivation to learn (93.3%), while the remaining participants declared to be skeptical (6.7%) (Figure 5a). Augmented Reality can be appropriately used to design treatment tools for children with dyslexia for 60% of the respondents, 33.3 % of them were skeptical while 6.7 % did not agree (Figure 5b).

As to the open question concerning how AR could be used in the design of treatment tools, two experts declared that AR could be used to create a more appealing interface, for example by using voices in rehabilitation tasks that are often boring and tiring for dyslexics, within an enriched context. Other respondents suggested that AR could provide reinforcements through multimodal channels and facilitate learning through more dynamic images (for instance, AR could support mathematical learning by directly providing the formulas to be applied or facilitating the visual representation of the problem), expanding the range of learning experience proposals or amplifying the stimuli to improve deficient functions and providing prompts for the identification of difficulties or errors.

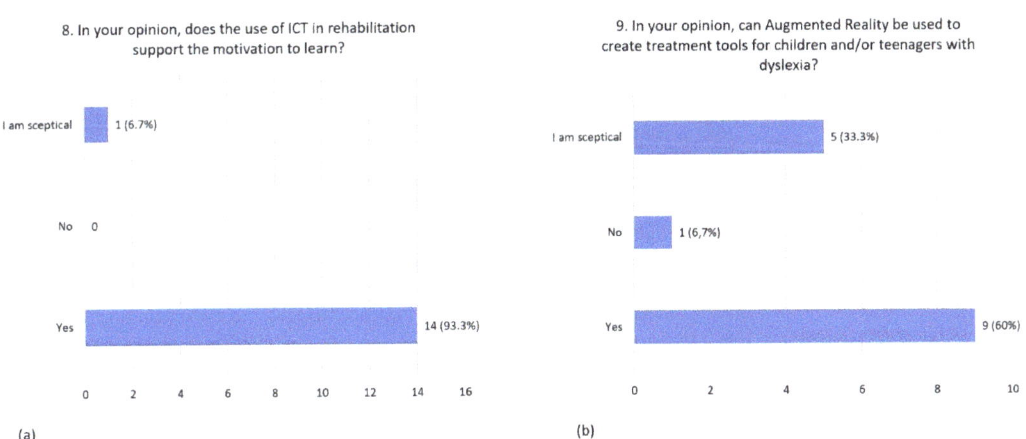

Figure 5. Distribution of responses to question 8 (**a**) and question 9 (**b**).

When asked to indicate from what age use of AR should be recommended, three experts replied that the ideal age is from 8 years on, three indicated the period of primary school (at the beginning or from the third grade), one respondent suggested use from 4 years on, one stated that AR could be used from the moment of diagnosis, and another suggested that the type of task should be taken into consideration. Other experts stated that among the aims of AR-based treatments could be the automation of metaphonological processes and global reading skills, the improvement of critical areas, the facilitation of the use of compensatory tools, the treatment of focused attention, and the shifting of attention, or more generally, to support learning and motivation ($n = 2$).

Virtual reality can be used to create intervention tools for children with dyslexia according to 60% of the respondents, 33.3 % of them were skeptical, while 6.7 % did not agree (Figure 6).

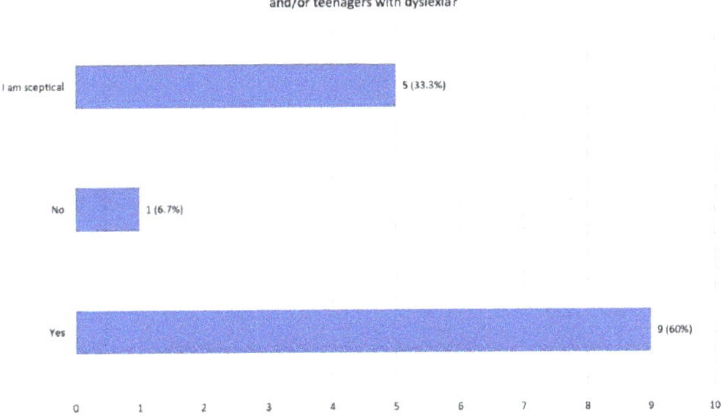

Figure 6. Distribution of responses to question 13.

Respondents stated that VR could be used to create learning environments to generalize acquired skills ($n = 1$) and to activate deficient skills through structured tasks in a playing situation ($n = 1$). Some experts suggested that VR could be used to enhance motor and spatial visual functions ($n = 1$); that it could be included in an integrated intervention program ($n = 1$), or in ecological contexts to facilitate learning through role playing activities ($n = 1$).

When asked from what age the use of VR could be recommended, the experts replied that the ideal age is from 8 years old or even before the age of 8 in subclinical or at-risk situations ($n = 3$). Other experts said that it could be used from primary school on ($n = 2$), from the moment of diagnosis, or depending on the type of task ($n = 2$).

Among the aims of the use of VR, the respondents listed increasing active participation and involvement ($n = 1$), activating deficient skills through exercises in the form of a game, enhancing learning, motivation and concentration, facilitating lexical access, attentional control, and perceptual discrimination.

As to the open question concerning the limitations in the use of ICT tools for the treatment of dyslexia, respondents said that is the tools are difficult to integrate into a global rehabilitation plan, may not be available at home, and require the involvement of the family if remotely operated. The level of satisfaction of the child, the risk of underutilizing the potential of digital technologies by merely proposing repetitive activities, the use of programs that engage the child through visual activities but do not really stimulate the decoding process, the possibility of feeding a dependency in subjects at risk, the reduction of social interactions and content sharing, economic issues, and the absence of mediation by the human expert (rehabilitator), were other reasons described by the experts.

ICT tools may facilitate the learning of school content in children with dyslexia according to 93.3% of respondents, while the remaining declared themselves to be skeptical (Figure 7).

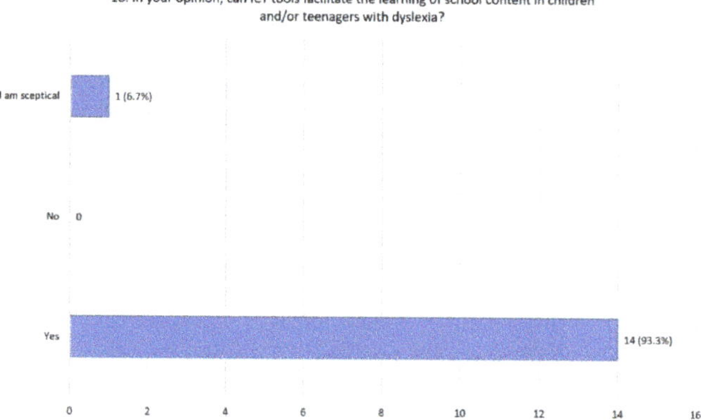

Figure 7. Distribution of responses to question 18.

When asked to imagine possible examples of ICT-based learning activities, experts offered different proposals, such as classes 3.0, promoting an online understanding of content research, the creation of study materials, the possibility of proposing the same multimedia content in different forms and with different degrees of complexity, encouraging creative and non-mnemonic learning, a different type of organization of activities, setting a time for a certain task, personal searches, and internet searches of study topics.

According to 53.3% of respondents, VR may be suitable for this purpose; 33.3% of them were skeptical, and 33.3% did not know (Figure 8a). Regarding AR, it may be suitable for this purpose for 53.3% of respondents, 20% of them were skeptical, and the remaining 26.7% did not know (Figure 8b).

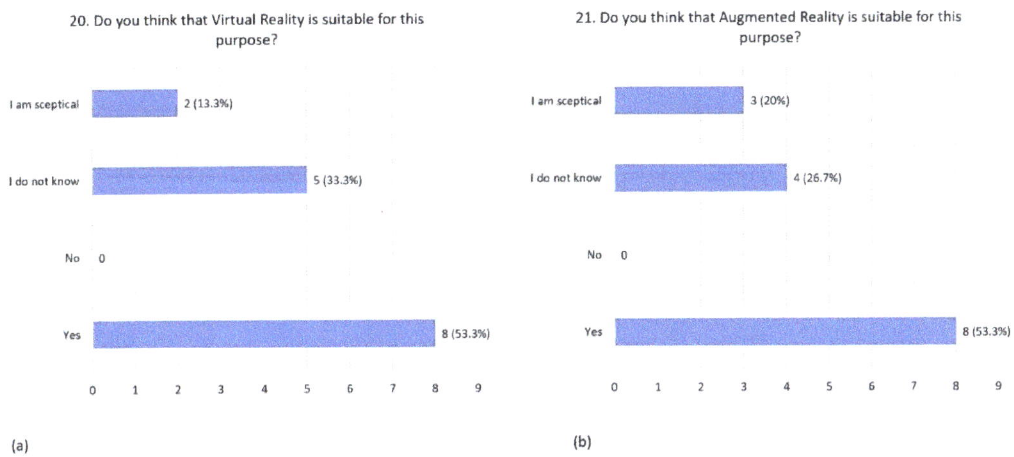

Figure 8. Distribution of responses to question 20 (**a**) and question 21 (**b**).

Based on the responses collected in Round 1, presented above, a list of statements was created and included in the questionnaire used in Round 2.

3.2. Round 2

During Round 2 we received responses from 15 of the 18 panel members (83.33%). Two of them sent the responses at the end of Round 2, when statement 6 and statement 8 of the second survey had already been modified for Round 3. For that reason, statements 6 and 8 have 13 answers while all the remaining have 15 answers.

The statements and the degree of agreement with the different questions are presented in Table 2.

Table 2. Round 2 statements and overall degree of agreement (expressing the percentage of "agree" and "strongly agree" replies on total number of responses excluding "I do not know" responses). Number of responses and percentages (in parentheses) are reported for each choice.

	Strongly Disagree *n* (%)	Disagree *n* (%)	Agree *n* (%)	Strongly Agree *n* (%)	I Do Not Know *n* (%)	Agreement (%)
(1) ICT technology can support the treatment of dyslexia as effectively as other methods can.	-	2 (13.33)	4 (26.67)	6 (40)	3 (20)	83.33
(2) ICT approaches can be seen as effective ways to integrate and, in some cases, substitute more traditional methods of treatment for developmental dyslexia.	-	-	7 (46.67)	6 (40)	2 (13.33)	89.27
(3) The main advantage of ICT approaches for the treatment of dyslexia is their flexibility, entailing the possibility to repeatedly propose the treatment several times per week, at the times that are more suitable for the children and their families.	-	2 (13.33)	2 (13.33)	9 (60)	2 (13.33)	87.69
(4) Other advantages of ICT trainings for dyslexia have to do with their ability to motivate and involve the children, and their ease of use. These characteristics allow children to work more with less effort.	-	1 (6.67)	1 (6.67)	6 (40)	7 (46.67)	90

Table 2. *Cont.*

	Strongly Disagree *n* (%)	Disagree *n* (%)	Agree *n* (%)	Strongly Agree *n* (%)	I Do Not Know *n* (%)	Agreement (%)
(5) Among the advantages of ICT trainings there is cost-effectiveness, although it is not considered to be a prominent factor for the choice of the training to be proposed.	1 (6.67)	-	4 (26.67)	6 (40)	4 (26.67)	85.45
(6) ICT training should primarily address the processes involved in assembling the phonological structure of the words.	1 (7.69)	3 (23.08)	4 (30.77)	3 (23.08)	2 (15.38)	69.09
(7) Other, secondary goals for ICT training for dyslexia should focus on the improvement of both visual analysis and lexical retrieval abilities.	1 (6.67)	-	5 (33.33)	7 (46.67)	2 (13.33)	86.15
(8) Grapheme-to-phoneme (and vice-versa) conversion processes may be involved in the ICT training, but they should not be considered as prominent goals of the intervention.	2 (15.38)	5 (38.46)	2 (15.38)	1 (7.69)	3 (23.08)	50
(9) The optimal duration of the training should be between 2 and 6 months.	-	3 (20)	7 (46.67)	5 (33.33)	-	78.67
(10) The ideal time for the start of training with ICT tools is from the third year of primary school. In some cases, the start can be anticipated to the first or second year of primary school.	-	3 (20)	7 (46.67)	4 (26.67)	1 (6.67)	77.14
(11) The use of ICT training can contribute to sustaining motivation for learning in general.	-	2 (13.33)	3 (20)	7 (46.67)	3 (20)	85
(12) Augmented reality can be employed in the design of ICT trainings for dyslexia, but it should not play a prominent role.	1 (6.67)	2 (13.33)	5 (33.33)	2 (13.33)	5 (33.33)	70
(13) Trainings based on augmented reality could be introduced starting from 7–8 years of age.	1 (6.67)	-	6 (40)	5 (33.33)	3 (20)	83.33
(14) Augmented reality could be used to enhance the salient characteristics of the stimuli to be processed.	-	-	5 (33.33)	7 (46.67)	3 (20)	91.67
(15) Augmented reality could be used to provide a multi-sensory, multi-modal environment during the tasks, enriching the quality and quantity of information regarding the stimuli.	-	-	6 (40)	6 (40)	3 (20)	90
(16) Augmented reality could be used to highlight the difficult aspects of the stimuli to be processed, so that the child is alerted and ready to activate and focus her/his resources during the task.	1 (6.67)	1 (6.67)	2 (13.33)	6 (40)	5 (33.33)	82
(17) Augmented reality could be used to provide additional information for specific stimuli, according to the needs and requests of the child.	-	-	6 (40)	6 (40)	3 (20)	90
(18) Augmented reality could be used to add motivating elements to repetitive, boring tasks to make them more appealing.	-	1 (6.67)	5 (33.33)	8 (53.33)	1 (6.67)	88.57

Table 2. *Cont.*

	Strongly Disagree *n* (%)	Disagree *n* (%)	Agree *n* (%)	Strongly Agree *n* (%)	I Do Not Know *n* (%)	Agreement (%)
(19) Augmented reality could facilitate automatization of metaphonological skills by highlighting processing units in the words (phonemes, syllables, and whole words).	-	2 (13.33)	5 (33.33)	4 (26.67)	4 (26.67)	80
(20) Further applications of augmented reality could be favoring attentional focusing and shifting processes.	-	-	7 (46.67)	4 (26.67)	3 (20)	83.33
(21) Additional applications of augmented reality in support of dyslexia extend to facilitating reading in everyday life contexts.	-	1 (6.67)	6 (40)	3 (20)	4 (26.67)	78.18
(22) Virtual reality can be employed in the design of ICT tools for the treatment of dyslexia.	-	2 (13.33)	7 (46.67)	4 (26.67)	1 (6.67)	77.14
(23) Training based on virtual reality could be introduced starting from 7–8 years of age.	-	1 (6.67)	6 (40)	5 (33.33)	3 (20)	85
(24) Virtual reality could be used to propose study subjects in realistic contexts, emphasizing the links between these subjects and real life.	-	-	5 (33.33)	6 (40)	4 (26.67)	90.91
(25) Virtual reality could be used to provide tasks embedded in ecologically plausible and varying contexts, thus fostering generalization processes.	-	-	6 (40)	6 (40)	3 (20)	90
(26) Virtual reality could be used to work on the child's difficulties in a structured way through engaging, motivating tasks, and games.	-	-	8 (53.33)	4 (26.67)	3 (20)	86.67
(27) Virtual reality could be used to train learned skills through simulations and role-playing activities.	-	-	6 (40)	6 (40)	3 (20)	90
(28) Virtual reality could be used to design integrated trainings involving reading as well as visual and motor functions simultaneously.	-	-	5 (33.33)	6 (40)	4 (26.67)	90.91
(29) Virtual reality could facilitate automatization of metaphonological skills, lexical access, and perceptual discrimination.	-	1 (6.67)	5 (33.33)	3 (20)	5 (33.33)	78
(30) Further applications of virtual reality could aim at improving attentional processes and executive functions.	-	-	7 (46.67)	5 (33.33)	2 (13.33)	84.62
(31) Additional applications of virtual reality could extend to training more effective management of negative emotions related to dyslexia and learning difficulties.	-	2 (13.33)	3 (20)	4 (26.67)	5 (33.33)	76
(32) While using ICT tools for the treatment of dyslexia, maximum attention should be devoted to avoiding the risk of addiction.	-	4 (26.67)	3 (20)	7 (46.67)	1 (6.67)	78.57

Table 2. *Cont.*

	Strongly Disagree n (%)	Disagree n (%)	Agree n (%)	Strongly Agree n (%)	I Do Not Know n (%)	Agreement (%)
(33) Use of ICT tools for dyslexia treatment should be proposed only after checking that adequate devices, connections, and familial support are available to the users.	-	-	2 (13.33)	13 (86.67)	-	97.33
(34) Use of ICT tools for the treatment of dyslexia should always be monitored by human supervisors who also ensure that the child's needs, opinions, and feelings are taken into account.	-	-	-	15 (100)	-	100
(35) Use of ICT tools should be designed as to provide activities that are not only engaging, but also meaningful for the children/teenagers with dyslexia.	-	-	2 (13.33)	12 (80)	1 (6.67)	97.14
(36) ICT tools, including virtual and augmented reality, can also be used to support learning of school contents in children/teenagers with dyslexia.	-	-	5 (33.33)	6 (40)	4 (26.67)	90.91
(37) Support of general content learning in students with dyslexia could be achieved through ad-hoc activities with increasing levels of difficulty and complexity, emphasizing real understanding, and assimilation of meanings.	-	-	4 (26.67)	9 (60)	2 (13.33)	93.85
(38) ICT tools for students with dyslexia could provide training for web-surfing and searching skills, and for creative, responsible use of internet sources and tools.	-	1 (6.67)	3 (20)	6 (40)	5 (33.33)	88
(39) ICT tools could support general learning in students with dyslexia by providing a series of ordered activities where organization of study materials is required, based on the integration of both (possibly facilitated) reading and other, multi-media sources of information.	-	1 (6.67)	4 (26.67)	7 (46.67)	3 (20)	88.33

Red figures indicate that the criterion of 75% agreement was not reached.

Experts provided ratings for each statement and qualitative data in the form of comments. There was a high level of agreement for most statements (mean 84.67%). Taking into account the requirement of 75% group consensus, all items achieved at least 76% agreement except for statement 6 (69.09%), statement 8 (50%), and statement 12 (70%).

Qualitative data on previous statements made it possible to understand the reasons for the low degree of agreement. Regarding statement 6 "ICT trainings should address primarily the processes involved in assembling the phonological structure of the words", experts who expressed a low level of agreement or gave a "I do not know" reply suggested that ICT trainings may address various processes involved in reading, not only the process involved in the phonological structure of the words.

Regarding statement 8 "Grapheme-to-phoneme (and vice-versa) conversion processes may be addressed in the ICT training, but they should not be considered as prominent goals of the intervention", three experts who expressed a low level of agreement argued that the grapheme-to-phoneme conversion process should be considered an important goal of the intervention, and another member of the panel specified that this goal depends on the child's age.

Lastly, statement 12 "Augmented reality can be employed in the design of ICT trainings for dyslexia, but it should not play a prominent role" received an uncertain reply from five panel members (33.33%) who declared "not to know" about the topic but did not add any comment or suggestion. For that reason, the statement was not modified for Round 3 and was finally deleted.

Statement 3 "The main advantage of ICT approaches for the treatment of dyslexia is its flexibility, entailing the possibility to repeatedly propose the treatment several times per week, at the times that are more suitable for the children and their families" received a high level of agreement (87.69%) and a comment regarding the importance of the quality of the intervention, so it was decided to add a new statement in Round 3 survey to understand more about the quality and adaptation of the intervention on the level of performance.

3.3. Round 3

On the basis of the comments provided by the panel to the statements of Round 2, we made some further modification to the survey. The revised set of modified statements was sent to the panel for further comment, and 11 experts gave their degree of agreement to the three new statements (statement 3b added on the basis of statement 3 comments, statement 6 and 8 to replace the previous ones). Table 3 presents the degree of agreement obtained regarding the three new statements (Table 3).

Table 3. Agreement ratings for the three statements added in Round 3 and the different ratings collected at Round 2 and Round 3. Number of responses and percentages (in parentheses) are reported for each choice.

		Strongly Disagree n (%)	Disagree n (%)	Agree n (%)	Strongly Agree n (%)	I Do Not Know n (%)	Agreement (%)
		Statement 6					
Round 2	ICT trainings should address primarily the processes involved in assembling the phonological structure of the words.	1 (7.69)	3 (23.08)	4 (30.77)	3 (23.08)	2 (15.38)	69.09
Round 3	ICT trainings may address the processes involved in assembling the phonological structure of the words.	0	1 (9.09)	4 (36.36)	3 (27.27)	3 (27.27)	82.5
		Statement 8					
Round 2	Grapheme-to-phoneme (and vice-versa) conversion processes may be involved in the ICT training, but they should not be considered as prominent goals of the intervention.	2 (15.38)	5 (38.46)	2 (15.38)	1 (7.69)	3 (23.08)	50
Round 3	Grapheme-to-phoneme (and vice-versa) conversion processes may be involved in the ICT training.	-	1 (9.09)	4 (36.36)	5 (45.45)	1 (9.09)	86
		Statement 3b					
Round 3	A further advantage linked to flexibility is the possibility to implement algorithms adapting the requests to the level of performance.	-	-	5 (45.45)	6 (54.55)	-	90.91

All items achieved at least 82.5% agreement. According to the results of Round 3, the final agreed version of the survey was made of 39 statements, 36 belonging to the survey of Round 2 and the three new statements of Round 3.

4. Discussion

The aim of the present study was to define a set of statements in order to form the basis for international "good practices" in the use of technologies for intervention and support for developmental dyslexia, in the context of the European "ForDys-Var" Erasmus+ project. The study was conducted using the Delphi method [31] reaching 75% minimum agreement on all of the statements after three rounds. Eighteen psychologists, child neuropsychiatrists and speech-and-language pathologists, among the Italian most recognized experts in developmental dyslexia, took part in the study, expressing their opinions on the topic through online questionnaires.

The first round of the survey consisted of 21 questions about technology applied to dyslexia. The second survey was realized starting from the quantitative and qualitative data obtained in Round 1 and consisted of 39 statements. The panel of experts expressed their degree of agreement with each statement providing answers on a Likert-type scale with four levels, from "strongly disagree" to "strongly agree". Two statements that did not reach 75% consensus were modified for the last survey based on the comments provided by the experts. Round 3 reached a 75% of consensus for all the statements, that were thus accepted as final recommendations.

Altogether, the recommendations emerging from the study indicate a favorable attitude by the panel members towards ICT-based intervention approaches for neurodevelopmental disorders in children, particularly dyslexia. Such practices are considered as generally effective and motivating. The experts, moreover, extended general clinical guidelines for the treatment of dyslexia to ICT treatment, recommending that it should be started before grade three and should last up to six months. Among the advantages of ICT-based approaches, the experts indicated flexibility, adaptivity (also in terms of self-adjusting algorithms), engagement, and cost-effectiveness. Among the specific advantages of AR, the experts underscored its capability to enhance specific stimulus characteristics as desired for the therapy, and its multi-sensory nature. Turning to VR, dyslexia experts appreciated its capability to create links between educational topics and real life, to sustain generalization processes, and to provide multi-sensory stimulation.

The functions listed as main targets for intervention include phonological awareness, visual abilities, lexical skills, and grapheme-to-phoneme (and vice versa) conversion.

In round 2, almost all the statements received a high level of agreement. Experts especially valued the effectiveness of ICT approaches in their integration to more traditional methods of treatment of dyslexia, increasing children's motivation and involvement. Moreover, great importance was given to the meaningfulness of the activities to be proposed (and their ecological validity) also through a multi-sensory, multi-modal environment that could enrich the quality and quantity of stimulus-related information. Great attention is also to be paid, according to the experts, to the families' compliance, to their possibility and capability to support the children during treatment, and to the fundamental role of constant human mediation and supervision during ICT-based activities.

Indeed, regular screen use is a fact of modern life: on average, children aged 8–12 in the United States spend 4–6 h a day using screens, while teens spend up to 9 h (data from American Academy of Child and Adolescent Psychiatry, February 2020) [37]. Media use guidelines around the world [38–40] encourage a correct use of screen and multimedia. For example, for preschool children, it is recommended to limit non-educational screen time to about 1 h per weekday and 3 h on weekend days; for school-age children, instead, no exact screen time limits are suggested, but guidelines highlight the importance to encourage healthy habits and limit activities that include screens. The absence of a precise cut-off for children's screen time depends on the lack of evidence on the differential effects of different forms of screen time [40]. However, it is considered appropriate to manage screen use, discouraging media use especially during homework or meals, while encouraging meaningful screen use. As for the present study and the applications of technology addressed in it, we believe that the use of ICT for intervention and support in developmental dyslexia, and its impact in educational settings, could be considered as a healthy (and meaningful) use of screen time, but close supervision should ensure that the use of such applications is not taking the form of addiction.

Among the limitations of the present study is the low level of expertise declared by many of the panel members with regard to clinical applications of virtual reality and, particularly, of augmented reality. Indeed, for some of the statements, the experts failed to provide an agreement response, with high percentages of "I do not know" replies. In fact, the experts had been identified as prominent scholars in their discipline and experts in the use of technology for the rehabilitation of developmental dyslexia, but they were not necessarily experts in the use of advanced technologies such as virtual reality and augmented reality. This confirmed that the use of ICT for the rehabilitation of reading disorders is, at least in Italy, almost exclusively limited to more traditional forms of technology such as computerized games and exercises, and possibly text-to-speech or speech-to-text applications to support school activities, whereas newer and more advanced technologies are still rarely known and used. Despite this, we believe that the panel was representative of the state of knowledge and expertise at a national level, and that similar (and possibly less informative) results would have been obtained by contacting a different group of professionals. While it would have been possible to include a panel of ICT experts with a more technical background, this would have implied lowering the required level of knowledge on the specific characteristics of learning disorders in children, which we believed to be a necessary requirement to be able to judge the impact and the effects of technology on the children's cognitive, psychological, and neuropsychological development.

In the context of the European "ForDys-Var" Erasmus+ project, the final set of statements will be sent to a more extended group of psychologists, child neuropsychiatrists, and speech-and-language pathologists, including teachers and school professionals of different European countries, in order to collect a consensus from a broader range of professionals and to define a set of recommendations and best practices to be shared at European level.

Author Contributions: Conceptualization, M.L.L.; methodology, M.L.L.; software, M.L.L.; validation, M.L.L. and F.B.; formal analysis, M.L.L. and F.B.; investigation, M.L.L. and M.D.R.; resources, A.M.; data curation, M.L.L. and F.B.; writing—original draft preparation, F.B., M.D.R., M.L.L.; writing—review and editing, M.L.L. and F.B.; visualization, F.B.; supervision, M.L.L. and A.M.; project administration, A.M.; funding acquisition, A.M. All authors have read and agreed to the published version of the manuscript.

Funding: This research was co-financed by the Erasmus+ program of the European Union through the 2018-1-ES01-KA201-050659 project. The work was further supported by the Italian Ministry of Health, Grant number RC2021 to M.L.L.

Institutional Review Board Statement: Data were collected as part of the project n. 902, approved by the Clinical Research Ethics Committee (CESC/CREC) of the provinces of Treviso and Belluno on 22 October 2020 (prot. 0172099) and fully authorized on 9 April 2021 as study prot. n. 823.

Informed Consent Statement: The data described were collected in a fully anonymous online survey. All respondents were informed that their answers would be collected in completely anonymous form and would be used for scientific purposes, and that by filling the questionnaires they accepted such conditions.

Data Availability Statement: The datasets have been deposited on Zenodo (doi:10.5281/zenodo.5572643) and are publicly accessible.

Acknowledgments: The authors wish to thank all the expert professionals who took part in the consultation.

Conflicts of Interest: The authors declare no conflict of interest. The funders had no role in the design of the study; in the collection, analyses, or interpretation of data; in the writing of the manuscript, or in the decision to publish the results.

References

1. Carter, M.; Grover, V. Me, my self, and I(T): Conceptualizing information technology identity and its implications. *MIS Q.* **2015**, *39*, 931–958. [CrossRef]
2. Gallucci, A.; Trimarchi, P.D.; Abbate, C.; Tuena, C.; Pedroli, E.; Lattanzio, F.; Giunco, F. ICT technologies as new promising tools for the managing of frailty: A systematic review. *Aging Clin. Exp. Res.* **2021**, *33*, 1453–1464. [CrossRef]
3. Beard, L.; Carpenter, L.; Johnston, L. *Assistive Technology: Access for All Students*, 2nd ed.; Merrill, Prentice-Hall: Upper Saddle River, NJ, USA, 2011.
4. Garzón, J. An Overview of Twenty-Five Years of Augmented Reality in Education. *Multimodal Technol. Interact.* **2021**, *5*, 37. [CrossRef]
5. Doğan, S.; Delialioğlu, O. A systematic review on use of technology in learning disabilities. *Ank. Univ. Fac. Educ. Sci. J. Spec. Educ.* **2020**, *21*, 611–638. [CrossRef]

6. Lerga, R.; Candrlic, S.; Jakupovic, A. A Review on Assistive Technologies for Students with Dyslexia. In Proceedings of the 13th International Conference on Computer Supported Education (CSEDU 2021), Prague, Czechia, 23–25 April 2021; Volume 2, pp. 64–72. [CrossRef]
7. Akçayır, M.; Akçayır, G. Advantages and challenges associated with augmented reality for education: A systematic review of the literature. *Educ. Res. Rev.* **2017**, *20*, 1–11. [CrossRef]
8. Cheng, K.H.; Tsai, C.C. Affordances of augmented reality in science learning: Suggestions for future research. *J. Sci. Educ. Technol.* **2013**, *22*, 449–462. [CrossRef]
9. Akçayır, M.; Akçayır, G.; Pektaş, H.M.; Ocak, M.A. Augmented reality in science laboratories: The effects of augmented reality on university students' laboratory skills and attitudes toward science laboratories. *Comput. Hum. Behav.* **2016**, *57*, 334–342. [CrossRef]
10. Tobar-Muñoz, H.; Baldiris, S.; Fabregat, R. Augmented reality game-based learning: Enriching students' experience during reading comprehension activities. *J. Educ. Comput. Res.* **2017**, *55*, 901–936. [CrossRef]
11. Rubin, P. *Future Presence: How Virtual Reality Is Changing Human Connection, Intimacy, and the Limits of Ordinary Life*; HarperCollins: New York, NY, USA, 2018.
12. Weiss, P.L.; Katz, N. The potential of virtual reality for rehabilitation. *J. Rehabil. Res. Dev.* **2004**, *41*, 7–10.
13. Bevilacqua, R.; Maranesi, E.; Riccardi, G.R.; Di Donna, V.; Pelliccioni, P.; Luzi, R.; Pelliccioni, G. Non-immersive virtual reality for rehabilitation of the older people: A systematic review into efficacy and effectiveness. *J. Clin. Med.* **2019**, *8*, 1882. [CrossRef]
14. Lorusso, M.L.; Travellini, S.; Giorgetti, M.; Negrini, P.; Reni, G.; Biffi, E. Semi-Immersive Virtual Reality as a Tool to Improve Cognitive and Social Abilities in Preschool Children. *Appl. Sci.* **2020**, *10*, 2948. [CrossRef]
15. Montana, J.I.; Tuena, C.; Serino, S.; Cipresso, P.; Riva, G. Neurorehabilitation of Spatial Memory Using Virtual Environments: A Systematic Review. *J. Clin. Med.* **2019**, *8*, 1516. [CrossRef] [PubMed]
16. Pedroli, E.; Padula, P.; Guala, A.; Meardi, M.T.; Riva, G.; Albani, G. A psychometric tool for a virtual reality rehabilitation approach for dyslexia. *Comput. Math. Methods Med.* **2017**, *2017*, 7048676. [CrossRef] [PubMed]
17. Jing, C.T.; Chen, C.J. A research review: How technology helps to improve the learning process of learners with dyslexia. *J. Cogn. Sci. Hum. Dev.* **2017**, *2*. [CrossRef]
18. Lyon, G.R.; Shaywitz, S.E.; Shaywitz, B.A. A definition of dyslexia. *Ann. Dyslexia* **2003**, *53*, 1–14. [CrossRef]
19. World Health Organization. *International Statistical Classification of Diseases and Related Health Problems—10th Revision*; World Health Organization: Geneva, Switzerland, 2011.
20. Cornoldi, C.; Tressoldi, P. Linee guida per la diagnosi dei profili di dislessia e disortografia previsti dalla legge 170: Invito a un dibattito [Guidelines for the diagnosis of dyslexia and dysorthography according to the law 170: Invitation to a debate]. *Psicol. Clin. Dello Svilupp.* **2014**, *18*, 75–92. [CrossRef]
21. Lorusso, M.L.; Vernice, M.; Dieterich, M.; Brizzolara, D.; Mariani, E.; Masi, S.D.; Mele, A. The process and criteria for diagnosing specific learning disorders: Indications from the Consensus Conference promoted by the Italian National Institute of Health. *Ann. Ist. Super. Sanit.* **2014**, *50*, 77–89. [CrossRef]
22. Peterson, R.L.; Pennington, B.F. Developmental dyslexia. *Annu. Rev. Clin. Psychol.* **2015**, *11*, 283–307. [CrossRef]
23. Zoccolotti, P. Consensus Conference su "La riabilitazione neuropsicologica della persona adulta" [Consensus Conference on Neuropsychological rehabilitation in adults]. *G. Ital. Di Psicol.* **2011**, *38*, 259–266.
24. Dziorny, M. Online course design elements to better meet the academic needs of students with dyslexia in higher education. In Proceedings of the Society for Information Technology & Teacher Education International Conference, Austin, TX, USA, 11 April 2012; Association for the Advancement of Computing in Education (AACE): Austin, TX, USA, 2012; pp. 332–337.
25. Bediou, B.; Adams, D.M.; Mayer, R.E.; Tipton, E.; Green, C.S.; Bavelier, D. Meta-analysis of action video game impact on perceptual, attentional, and cognitive skills. *Psychol. Bull.* **2018**, *144*, 77. [CrossRef] [PubMed]
26. Bertoni, S.; Franceschini, S.; Puccio, G.; Mancarella, M.; Gori, S.; Facoetti, A. Action Video Games Enhance Attentional Control and Phonological Decoding in Children with Developmental Dyslexia. *Brain Sci.* **2021**, *11*, 171. [CrossRef]
27. Chopin, A.; Bediou, B.; Bavelier, D. Altering perception: The case of action video gaming. *Curr. Opin. Psychol.* **2019**, *29*, 168–173. [CrossRef]
28. Franceschini, S.; Trevisan, P.; Ronconi, L.; Bertoni, S.; Colmar, S.; Double, K.; Facoetti, A.; Gori, S. Action video games improve reading abilities and visual-to-auditory attentional shifting in English-speaking children with dyslexia. *Sci. Rep.* **2017**, *7*, 5863. [CrossRef] [PubMed]
29. Peters, J.L.; Crewther, S.G.; Murphy, M.J.; Bavin, E.L. Action Video Game Training improves Text Reading Accuracy, Rate and Comprehension in Children with Dyslexia: A Randomized Controlled Trial. *Sci. Rep.* **2021**, *11*, 18584. [CrossRef]
30. Cidrim, L.; Madeiro, F. Information and Communication Technology (ICT) applied to dyslexia: Literature review. *Rev. CEFAC* **2017**, *19*, 99–108. [CrossRef]
31. Hasson, F.; Keeney, S.; McKenna, H. Research guidelines for the Delphi survey technique. *J. Adv. Nurs.* **2000**, *32*, 1008–1015. [CrossRef]
32. Von der Gracht, H.A. Consensus measurement in Delphi studies: Review and implications for future quality assurance. *Technol. Forecast. Soc. Chang.* **2012**, *79*, 1525–1536. [CrossRef]
33. Boulkedid, R.; Abdoul, H.; Loustau, M.; Sibony, O.; Alberti, C. Using and reporting the Delphi Method for selecting healthcare quality indicators: A systematic review. *PLoS ONE* **2011**, *6*, e20476. [CrossRef]

34. Shinners, L.; Aggar, C.; Grace, S.; Smith, S. Exploring healthcare professionals' perceptions of artificial intelligence: Validating a questionnaire using the e-Delphi method. *Digit. Health* **2021**, *7*. [CrossRef] [PubMed]
35. Murphy, M.K.; Black, N.A.; Lamping, D.L.; McKee, C.M.; Sanderson, C.F.; Askham, J.; Marteau, T. Consensus development methods, and their use in clinical guideline development. *Health Technol. Assess.* **1998**, *2*, 1–88. [CrossRef]
36. Gill, F.J.; Leslie, G.D.; Grech, C.; Latour, J.M. Using a web-based survey tool to undertake a Delphi study: Application for nurse education research. *Nurse Educ. Today* **2013**, *33*, 1322–1328. [CrossRef] [PubMed]
37. Available online: https://www.aacap.org/AACAP/Families_and_Youth/Facts_for_Families/FFF-Guide/Children-And-Watching-TV-054.aspx (accessed on 5 November 2021).
38. Canadian Paediatric Society, Digital Health Task Force, Ottawa, Ontario. Digital media: Promoting healthy screen use in school-aged children and adolescents. *Paediatr. Child Health* **2019**, *24*, 402–408. [CrossRef] [PubMed]
39. World Health Organization. *Guidelines on Physical Activity, Sedentary Behaviour and Sleep for Children under 5 Years of Age*; World Health Organization: Geneva, Switzerland, 2019.
40. Viner, R.; Davie, M.; Firth, A. *The Health Impacts of Screen Time: A Guide for Clinicians and Parents*; Royal College of Paediatrics and Child Health: London, UK, 2019.

MDPI
St. Alban-Anlage 66
4052 Basel
Switzerland
Tel. +41 61 683 77 34
Fax +41 61 302 89 18
www.mdpi.com

Children Editorial Office
E-mail: children@mdpi.com
www.mdpi.com/journal/children

www.ingramcontent.com/pod-product-compliance
Lightning Source LLC
LaVergne TN
LVHW070611170726
843515LV00004B/42

* 9 7 8 3 0 3 6 5 4 2 9 2 8 *